BOTANICAL RESEARCH AND PRACTICES

MACROPHYTES

BIODIVERSITY, ROLE IN AQUATIC ECOSYSTEMS AND MANAGEMENT STRATEGIES

BOTANICAL RESEARCH AND PRACTICES

Additional books in this series can be found on Nova's website under the Series tab.

Additional e-books in this series can be found on Nova's website under the e-book tab.

BOTANICAL RESEARCH AND PRACTICES

MACROPHYTES

BIODIVERSITY, ROLE IN AQUATIC ECOSYSTEMS AND MANAGEMENT STRATEGIES

RAVEN CAPELLO
EDITOR

NOVINKA

New York

For permission to use material from this book please contact us:
Telephone 631-231-7269; Fax 631-231-8175
Web Site: http://www.novapublishers.com

LIBRARY OF CONGRESS CATALOGING-IN-PUBLICATION DATA

Macrophytes : biodiversity, role in aquatic ecosystems and management strategies / editor: Raven Capello.
pages cm. -- (Botanical research and practices)
Includes index.
ISBN 978-1-63321-518-4 (softcover)
1. Aquatic plants. 2. Aquatic ecology. I. Capello, Raven. II. Series: Botanical research and practices.
QK930.M33 2014
581.7'6--dc23

2014026716

Published by Nova Science Publishers, Inc. † New York

CONTENTS

PREFACE

Aquatic plants play a significant role in maintaining water quality, due to their ability of absorbing nutrients. This book aims to show the value of macrophyte use as a tool for biological water treatment of aquaculture farms and also as a culture medium for microalgae. It also discusses the translocation and monitoring of wetland plants; responses of floating macrophytes to different environmental stressors; and vegetation structure of lagoon environments as a result of sediment biogeochemical dynamics and organic matter load.

Chapter 1 – Wet habitats are among the most threatened ecosystems all over the World and the importance of wetland protection is well recognized in several international treaties. Threats for wetlands are numerous; drainage in agriculture, use of water for the industry and agriculture, climate change-mediated drought or intrusion of salt water into freshwater wetlands, water pollution and eutrophication. Consequently, wetland destruction, fragmentation and degradation have led to negative direct and indirect impacts on wetland-dependent plants. Habitat protection, restoration or reconstructions are widely used ways to recover aquatic ecosystems.

Habitat restoration and reconstruction require wetland plants to be translocated into restored sites or new sites. However, translocation of wetland plants can be very challenging because it also includes the water quality management. In many times, this is the most difficult part in translocation programs. Water quality can be assessed through the analysis of several physical and chemical variables and long time series of sampling is needed to obtain reliable data on water quality. In fact, while soil accumulates pollutants and nutrients, the concentration of pollutants in running water may reach high values during acute and short-lasting episodes, which makes them difficult to be detected. In some cases, also toxicological analysis may be needed to understand the thresholds of specific factors affecting the survival of the target species.

In general, translocations require sound knowledge on species and habitats, like population dynamics, mating system and ability of dispersal. The selection and restoration of the target sites are one of the most important phases in a translocation program. Although different techniques are available to avoid mistakes in choosing the target sites, it must be considered that their selection should be quick and cheap. Here the authors present different approaches for the selection of suitable translocation sites for wetland-dependent plants species.

Although several guidelines to translocation have been recently developed, few specific best practices for wetland-dependent plants are available. Thus, here the authors provide some practical advices and examples useful to improve the translocation success of wetland plants, from the preliminary phases of translocation to the monitoring program.

Chapter 2 – During their life plants have to cope with natural or anthropogenic environmental stressors, including different pollutants. The exposure of aquatic macrophytes to these compounds mobilizes or generates, directly or indirectly, diverse signalling molecules and regulates many processes that amplify and specify the physiological response through transcriptional and metabolic changes. Due to consistent enhancement of water pollution and in order to better preserve aquatic ecosystems, the knowledge of response mechanisms used by the plants to cope with stress is of increasing significance. In this paper, the stressor driven changes and responses induced by different pollutants and phytoremediation potential of some floating macrophyte species will be considered and discussed.

Chapter 3 – Shallow lagoons with low water renewal tend to be eutrophic. Eutrophic conditions may become extreme if there is strong human pressure in the surrounding area. They promote opportunistic macroalgae which can constitute vast dense mats. Subsequent decay of biomass enriches sediment with organic matter and may lead to dystrophic status. During these phases, mineralization processes involving bacterial activities, such as ammonification, nitrification, denitrification and sulphate reduction, occur in sediment. Decay and mineralization of biomass imply release of toxic catabolites and nutrients, the latter supporting new blooms of vegetation. Under extreme conditions, macroalgal blooms repeat with increasing frequency, culminating in a dissipating event, which manifests as an extended dystrophic crisis. Here the authors report a conceptual model of the sediment s can be very challenging because it also includes the water gement. In many times, this is the most difficult part in programs. Water quality can be assessed through the veral physical and chemical variables and long time series s needed to obtain reliable data on water quality. In fact, cumulates pollutants and nutrients, the concentration of unning water may reach high values during acute and short-les, which makes them difficult to be detected. In some oxicological analysis may be needed to understand the specific factors affecting the survival of the target species.

ıl, translocations require sound knowledge on species and population dynamics, mating system and ability of selection and restoration of the target sites are one of the nt phases in a translocation program. Although different e available to avoid mistakes in choosing the target sites, it idered that their selection should be quick and cheap. Here fferent approaches for the selection of suitable translocation nd-dependent plants species.

several guidelines to translocation have been recently w specific best practices for wetland-dependent plants are us, here we provide some practical advices and examples rove the translocation success of wetland plants, from the hases of translocation to the monitoring program.

INTRODUCTION

slocation?

ɔ [10], "translocation is the human-mediated movement of from one area, with release in another". However, the slocations only include those actions that produce benefit for ɔcated population or the recipient ecosystem. In this sense, n-native (or invasive) species are not included in the concept translocation. The conservation translocations include with different purposes: a) population reinforcement is the uals to an existing population with the purpose to increase the b) reintroduction is the establishment of a new population in e historical native range of the target species, from which it c) ecological replacement is the release of a species with l or functional characteristics of an extinct species and it is ablish the ecological function of an ecosystem. More

energy cycle, from low organic matter content of sediment to accumulation and anoxigenic dissipation, and thence to new accumulation, and so on. At the various stages, different quantities and qualities of nutrients are released from the sediment, creating different environmental conditions and blooms of vegetation that differ in quantity and quality.

Chapter 4 – Aquatic plants play a significant role in maintaining water quality, due to their ability of absorbing nutrients. The aim of this chapter is to show the value of macrophyte use as a tool for biological water treatment of aquaculture farms and also as a culture medium for microalgae. In southeastern Brazil, aquaculture farming is mainly carried out in flow-through systems, in which large earthen fishponds are commonly arranged in line, each pond receiving the effluent from the previous one, without any prior treatment. The continuous water flow through the fishponds strongly depends on seasonal rains. The current system for production of aquatic organisms could benefit greatly from the development of low cost strategies to mitigate its impacts in the environment. Maintenance of good water quality is of primary importance to maintain an ideal culture ecosystem, with adequate feed for optimal fish yield potential, and preservation of a eutrophic situation. Fish production may cause serious environmental issues, especially when there is little concern over excessive plant growth, which can influence the functioning and structure of ecological processes in the fishponds. The use of macrophytes as bio-filter and / or wetland or as microalgae culture medium in aquaculture farms can serve as a tool for maintenance of good environmental conditions and improve the cultivation of microalgae. These microalgae can in turn be used as a nutritional source for the fish larvae, pigment in aquaculture and also as a supplementary additive in the food industry. This alternative medium should allow efficient production of microalgae, at a low cost for the laboratory. The understanding of local characteristics and development of proper feeding techniques, aimed at the integration and optimization of fish production, are important factors in obtaining and maintaining water quality appropriate for microalgae culture.

In: Macrophytes
Editor: Raven Capello

Chapter 1

TRANSLOCATION
OF WETLA

Thomas Abeli[1,3,4], Grazi
and Simon
[1]University of Pavia, Department o
Pavi
[2]University of Pavia, Departme
Pavi
[3]IUCN/SSC Reintrod
[4]IUCN/SSC Freshwat

ABS

Wet habitats are among the m
World and the importance of wet
several international treaties. Threa
in agriculture, use of water for
change-mediated drought or intr
wetlands, water pollution and e
destruction, fragmentation and de
and indirect impacts on wetland-
restoration or reconstructions are
ecosystems.

Habitat restoration and recon
translocated into restored sites or

wetland p
quality m
translocati
analysis o
of sampli
while soi
pollutants
lasting ep
cases, als
thresholds

In ge
habitats,
dispersal.
most impo
techniques
must be co
we present
sites for we

Althou
developed,
available.
useful to i
preliminary

What Is Tra

According
living organis
conservation tr
the native tran
movements of
of conservatio
different actior
release of indiv
population size
an area within
has disappeare
similar ecologi
aimed to re-e

controversial is assisted colonization, meaning the release of species outside their native range to avoid extinction due to current or future projected threats (e.g. climate change).

Need for Wetland Plant Conservation

Wetland plants are important for aquatic ecosystems, because they perform several ecological functions. Aquatic plants are the main source of oxygen in water bodies, are the direct or indirect source of food for several organisms (fishes, waterfowls, etc.), provide shelter for fishes, amphibians and other aquatic animals, affect the nutrient cycle of both water bodies and banks [1] and in some cases, may represent a habitat them-selves. For instance, water soldier (i.e. *Stratiotes aloides* L.) vegetation may provide shelter for tens of invertebrate species, including the endangered dragonfly *Aeshna viridis* Eversmann, the reproduction of which is strictly related to the water soldier [2]. *S. aloides* also contributes to the reproductive success of the Black Tern [3]. Moreover, wetland plants provide several ecosystem services that may reflect on human well-being [4]. The most immediate function of aquatic plants is the regulation of the water flow, and riparian vegetation also provides protection against erosion. The importance of aquatic plants and their habitats (rivers, lakes, ponds, etc.) has been recognized since 1971 by the Ramsar Convention, a treaty for the international cooperation on wetland conservation (Figure 1).

Figure 1. Natural wetlands in the Danube Delta, Romania – Photo S. Orsenigo.

The conservation of wetland-dependent plants is the recurrent issue in conservation biology worldwide. However, the current conservation status of aquatic plants is shockingly poor, especially in the global perspective. In Europe, about 17% of the wetland species assessed against the IUCN criteria are threatened by extinction, 19% in Africa with one species declared extinct in the wild (*Nymphaea thermarum* Eb. Fisch.), 10% in South America, 8% in Asia and North America, 3% in Oceania (data extracted from the IUCN Global Red List at www.iucnredlist.org, accessed on April 2014). According to [5], approx. 7% of the wetland species legally protected by the European Union at the continent level are threatened, 1.3% being Critically Endangered. In the Mediterranean Basin, about 15% of wetland plant species are threatened, but when considering the endemic species in the Mediterranean Biodiversity hotspot, about 37% of the species are at the risk of global extinction [6, 7].

Increasing Need for Translocations

Given the poor conservation status of a high proportion of wetland plant species, conservation actions are necessary. Legal protection is not always effective in removing the threats affecting rare and endangered species and in many cases does not allow species that have disappeared from historical sites to re-establish viable populations. So, the recovery of extremely rare or disappeared species within their original range is often possible only by the human intervention.

As said above, translocations may help to restore species subjected to historic decline and to protect endangered taxa. Importantly, the protection or restoration of an ecosystem as a whole is likely to be the best conservation practice, because it assures the correct way for ecosystem functioning, including the complex trophic chains and inter-specific relationships. This is especially important for running water habitats, like rivers, streams or interconnected water bodies, where the water quality may be influenced by upstream conditions even at great distances. This is the reason why translocations, especially of species living in or nearby the water, should be included in broad projects of habitat restoration that assure suitable and stable water quality conditions on a large spatial scale.

In the very critical cases, assisted colonization provides the opportunity for rare species that will likely be affected by some threats in the future (e.g. planned infrastructures, climate change) to survive. This type of action was especially conceived to mitigate the effect of climate change, but its real effectiveness is controversial compared to the intrinsic risks of releasing species outside their native range [8].

Consequently, translocations may represent an option for plant conservation, but not an easy one. The rate of success is often very low [9]. There are several limitations in applying translocation techniques, often due to poor knowledge of the ecological requirements of the target species or the lack of propagules. General guidelines to translocations are available [10, 11, 12], but without any specific approach on wetland plants. In this chapter, we highlight challenges, propose solutions for general problems and identify best practices for wetland plant translocations based on the review of successful and unsuccessful translocation attempts worldwide and on authors' personal experience on translocation of wetland plants species. For information and techniques concerning plant translocation in general we recommend the existing literature and guidelines [e.g. 10, 11, 12].

Lack of Practice in Translocations of Wetland Plants

Plant translocations have shorter history than animal translocations, in particular when compared to vertebrates [13]. Consequently, the advancing of knowledge and development of appropriate techniques for translocations in plants, is rather poor. It follows, that there is a substantial lack of practice in translocations of wetland plants with the specific conservation purpose. This is due to a very low number of wetland plants included in translocation attempts (S. Godefroid, personal communication), but also to the fact that information on plant translocations is often not available in the scientific literature. Many successful or unsuccessful trials in wetland plant translocations are written to internal reports or are not published.

Some information is possible to gather from other disciplines that require plant releases like phyto-remediation. In fact, several riparian and aquatic plant species are successfully used for the phyto-remediation, like *Phragmites australis* (Cav.) Trin. ex Steud., *Iris pseudacorus* L., *Typha* spp. Other species used are common ornamental or food plants (e.g. the duck potato of the indigenous people of the Americas *Sagittaria latifolia* Willd.), which in may become invasive when released outside of their native range (e.g. the water lettuce *Pistia stratiotes* L. or *Eichhornia crassipes* (Mart.) Solms.). However, it must be noted that there is a substantial difference between the conservation translocation of wetland plants and the extensive use of some species for other purposes. Species requiring conservation translocations are usually rare or endangered because of persisting threats or reduced reproductive performance or very slow growth rate; instead, species used for phyto-remediation are chosen because of their broad ecological requirements and their easy propagation.

METHODS OF TRANSLOCATION OF WETLAND-DEPENDENT PLANT SPECIES

The Aim of Translocations Must be to Establish Self-Sustaining Populations Otherwise Better to Maintain Genetic Resource *Ex Situ*

The main aim of conservation translocations is to provide the benefit for the translocated population. Such benefit may be measured in different ways depending on the specific scopes of the translocation project. Generally, translocations aim to establish viable and self-sustaining populations, which are able to reproduce and to naturally follow their evolutionary and adaptive processes.

The meta-population approach seems to have the best response to this necessity and it is preferable to establish many populations i.e. subpopulations connected each other by gene flow (exchange of pollen, seeds or other propagules) [14]. The establishment of meta-populations reduces the risk of failure due to stochastic events (demographic, genetic, human-mediated), which rarely will affect all subpopulations at the same time.

One advantage of translocations in aquatic or riparian species is that most of them have hydrochorus dispersal by fruits (e.g. *Cakile edentula* (Bigel) Hook), seeds/spores (e.g. *Nyssa aquatica* L. /*Isoëtes* spp.) or vegetative propagules like turions (e.g. *Stratiotes aloides* L.) and rhizomes (e.g. *Nuphar luteum* (L.) Sm.). This helps to maintain the gene flow between populations even at great distance, which is advantageous for two reasons: 1) it may be easier to enhance gene flow between translocated-translocated populations or between translocated–natural populations, with beneficial effects on the genetic diversity and meta-population dynamics [14, 15] and 2) the source populations can be selected even from the great distance with a low risk of outbreeding depression. However, there are some cases where the gene flow follows different patterns [16] and where the genetic diversity of possible source populations should be studied beforehand. In general, populations close to their translocation sites (for reintroduction and reinforcement) and with high genetic variability should be preferred as the source population of plant material. Typically a high genetic variability increases the probability of successful establishment. However, depending on reproductive biology and colonization patterns of the target species, the closest populations may not always be the best source populations (authors' observation; *S. aloides*), and investigation of population genetic diversity and gene flow is recommended.

Beyond the establishment of the viable population of the target species, translocations may also have other objectives,, like the restoration of an ecosystem function or increasing the public awareness on plant conservation [17]. Plant translocations should be carried out by taking care of the natural dynamics of the target habitats and in compliance with the other occurring species. Moreover, it should be taken into account that rare and endangered species are often legally protected, so their translocation may require special permissions (e.g. CITES, national legislation, etc.).

In a preliminary phase of a translocation program, it is important to evaluate the feasibility of the project. In addition, it should be noted that, sometimes other conservation solutions might be more effective, like *ex situ* conservation (e.g. seed banking, nursery, botanical garden collection, etc.). In some cases, this would save time needed to ensure the survival of the target species.

The Role of *Ex Situ* Seed Banks and Cultivation

Ex situ conservation may represent the alternative way to translocation when the lack of information on the target species, the lack of suitable receptor sites or the unfeasibility to remove the threats affecting the target species suggest that the translocation may fail. Nevertheless, *ex situ* collections may help to solve one of the major problems in translocation: to obtain enough plant material for viable populations, especially with very rare species, trees and slow growing species (e.g. isoëtids). It is impossible to say the right number of individuals, because it is species-specific, but it is proven that the higher the number of plants released the higher the probability of success [9]. An important role in plant production for translocation is certainly played by *ex situ* plant collections both as seeds (seed banks) and living plants (botanical gardens, nurseries, herbaria; [19, 20]). The use of *ex situ* material as a source for translocation prevents damages for the natural populations (Figure 2). Of course, in the case of species extinction in the wild the *ex situ* material is the only choice. In these cases, seed banks and botanical gardens will become important centres for *ex situ* propagation and cultivation of the target species for translocation [21]. Nevertheless, the use of *ex situ* source populations, do not guarantee a high genetic variability and it should be considered that many aquatic plants have recalcitrant seeds, so they cannot be desiccated and stored in seed banks [18]. If *ex situ* material of the target species is limited, the collection of seeds/spores from natural populations will have smaller impact, than the collection of mature individuals.

Recently, also some private companies have become specialized in the propagation of wild plant species and have developed appropriate techniques to propagate the target species for translocations (BOX 1).

Figure 2. *Ex situ* propagation of *Marsilea quadrifolia* – Photo P. Cauzzi.

Box 1. An Example of Private Company that Produce Native Flora: Flora Conservation S.R.L.

One of the major problem of translocation is the availability of enough plant material to transplant, especially with rare and slow growing species. It is clearly demonstrated that higher the number of plants released in the receptor site higher the probability of success of the translocation. Institutions like botanical gardens or seed banks host *ex situ* collections of germplasm that can be used as a source material for propagation.

However, such institutions involved in translocation projects, may not have adequate facilities to produce in a short time a sufficient number of propagules to be translocated. In the last few years, several private companies specialized on native seeds and wild plants have developed, especially in Europe and North America, may help practitioners to overcome the problem of propagule availability. Flora Conservation s.r.l. is an example of a private company led by a group of Italian botanists and plant ecologists, with a sound experience in translocation, the mission of which is to produce native seeds and plant species (especially rare and endangered species) to be used in restoration projects (e.g. phytoremediation, translocation).

Plants are reproduced from seeds collected from wild populations, minimizing the pressure on wild populations and maximizing the genetic diversity of *ex situ* collections. Flora Conservation is specialized on common species used for phytoremediation (e.g.: *Carex acutiformis* Ehrh., *Phragmites australis* (Cav.) Trin. ex Steud., *Typha latifolia* L.), ecological restoration (e.g.: *Butomus umbellatus* L., *Iris pseudacorus* L., *Iris sibirica* L., *Lythrum salicaria* L.) and rare or threatened species subject of translocation projects (e.g.: *Jacobaea paludosa* subsp. *angustifolia* (Holub) B. Nord. & Greuter, *Nymphaea alba* L.).

Fundamental Need to Collect Information on the Ecology of the Species and of the Characteristics of the Receptor Site

In general, translocations require deep understanding of ecological requirements of the target species and whether the receptor sites meet the ecology of the species. A complete identification and description of the existing sites of the target species may provide detailed information on its ecology, population dynamics and threats. Additionally, a complete survey of the target species distribution may lead to a rethinking of the need of translocation.

Taxonomy, ecology, distribution, demography, reproductive biology and genetics of the target species are aspects that should be considered [22] (Figure 3). The lack of knowledge in one of these factors may prevent successful translocation. Furthermore, it is important to develop appropriate cultivation techniques that will enable to obtain the plant material for translocations.

Figure 3. Investigating the ecology of *Isoëtes malinverniana* before translocation – Photo T. Abeli.

When planning translocations of a wetland-dependent species it is first necessary to notice that "wetland" includes a variety of ecosystems, which each are diverse for several aspects. Consequently, translocation techniques between ecosystems may differ as well. For example, wetlands can be classified on the basis of the water flow (i.e. river or lake), the nutrient load (i.e. oligotrophic, mesotrophic, eutrophic), the water availability (i.e. permanent water bodies or temporary ponds), the water level (i.e. deep or shallow) and the salinity (i.e. freshwater or brackish). All these characteristics can co-exist within a single ecosystem and influence the plant community composition, the species ecological requirements and, finally, the translocation techniques (e.g. methods of ecological investigation, preparation of the receptor site, gene flow management, etc.).

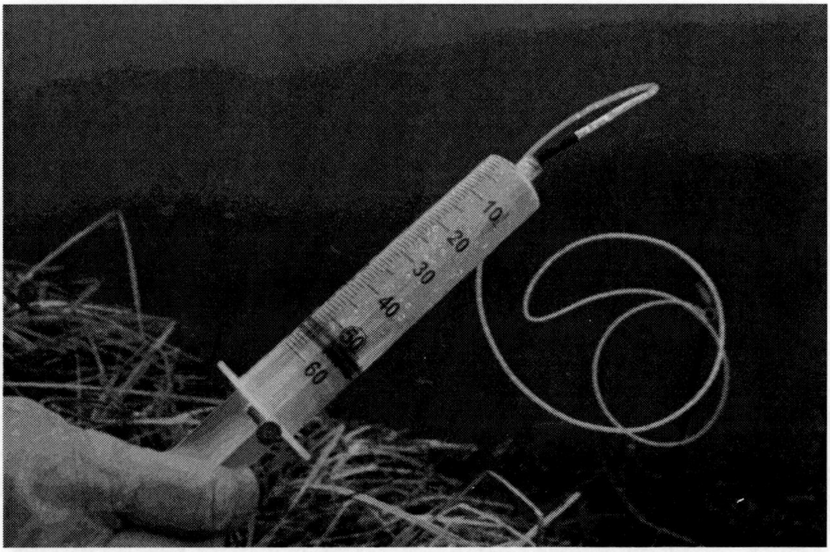

Figure 4. Sampling procedure of sediment pore water – Photo T. Abeli.

Secondly, wetland plants are strongly related to both the water and the soil (with the exception of floating species), which represents the main difference (and challenge) between the translocation of terrestrial plants and wetland-dependent species. In the latter case, greater attention should be paid to water or soil or both. While changes in soil chemistry are usually slow, flow, physical and chemical parameters of the water may periodically vary within relatively short periods (days, seasons) or may be affected by periodical episodes (e.g. flooding, pollution, etc.). This will create challenges for the studies of the target species and the suitability of sites.

A third variable, often neglected, is the hydrochemical characteristics of the sediment pore water (Figure 4), which may differ from both soil and water hydrochemistry, as revealed by studies on ecological requirements of different aquatic species [23, 24]. Moreover, translocation techniques vary depending on whether the target species is an aquatic or riparian species.

Need to Identify the Reason Why the Taxon Has Declined or Been Lost from the Site

A general recommendation in translocation is the removal of existing and potential threats for the target species. Translocations, especially reintroductions, often fail because the threats that brought a species to local extinction were not properly removed or mitigated [9]. Moreover, it has been demonstrated that reintroductions in protected areas have the higher success rate. Understanding the threats affecting the target species includes the investigation of direct effects, like population destruction due to human development or active predation and indirect effects like disruption of fundamental ecological functions, which in turn may affect the target species.

Within certain limits it is important to predict the possibility of future threats (e.g. development of infrastructure like dams) or stochastic events (e.g. climate fluctuation) that may affect the translocated populations. In addition, a receptor site may be influenced by other events from surrounding area. Change in water regime or discharge of pollutants may affect water quality at great distance. Only the guess of negative impacts alone, may give the justification for translocation, if the population will predicted to be lost through future action (e.g. development), if the factor suppressing the population (i.e. causing rarity) has ceased but the taxon has not recovered (e.g. peregrine banning of organochlorine pesticides), if insufficient dispersal is the factor stopping the taxon from spreading (e.g. *Ranunculus ophioglossifolius* Vill.), or if it is no longer possible to manage the site in a way that will enable the survival of the species, but this could be done at other sites (e.g. *Damasonium alisma* Mill. in UK).

Methods of Translocation

One way of translocations is direct translocation from the soil seed-bank or by released seeds. The other way is to use *ex situ* cultivation followed by re-introduction of seedlings or sub-adult and adult fruiting plants.

In general, the first way should be avoided, because the release of seeds strongly reduces the success rate of translocation, as seedlings are much more sensitive to environmental stress than older plants [9, 25]. Moreover, to guarantee the certain probability of success, a huge amount of seeds will be required. This option may be used for riparian species, but it is totally inapplicable for strictly aquatic plants, as the seeds will flow away from the target site. The second way guarantees higher success of population establishment, especially when used propagules of different ages (e.g. seedlings, sub-adults and adult fruiting individuals; Figure 5). The distribution of the age structure can be inferred from the study of natural populations, if available.

Figure 5. Reintroduction of *Leucojum aestivum*. Adult and sub-adult plants placed in the same plot – Photo P. Cauzzi.

The Use of Artificial Islands

It is important to know that translocation of wetland-dependent species is the most successful if they are performed in the context of broader habitat restoration projects.

Figure 6. An old quarry along the River Po (Italy) becomes slowly to a new wetland – Photo P. Cauzzi.

Figure 7. Planting plant individuals on artificial floating islands – Photo P. Cauzzi.

In particular, wetland plant translocations fit very well in the restoration of abandoned stone, gravel or sand quarries along rivers [26] (Figure 6). In such cases, translocations may represent an opportunity for population reinforcement of rare species and it is the only way, which allows native species to establish in areas, which are often highly "attractive" for alien species. However, restoration of dismissed quarries often requires bottom and riparian reshaping in order to create suitable microhabitats for plants [26]. When reshaping is not possible or when the water level has high variation, artificial floating islands may be a solution in lakes or ponds that have not suitable habitats for riparian or aquatic species (quarries are often too deep). Such islands consist of a floating structure, which can also be biodegradable, supporting a planter with holes (Figure 7). For now, they are produced by few companies, and their use may be expensive. However, current experiences have yielded excellent results in restoration and phytoremediation projects, especially because they adapt to the water level fluctuation [27, 28].

Selection of Receptor Sites

The choice of suitable translocation sites is the first crucial operative problem to deal with in a translocation program. Despite the importance of the selection of suitable receptor sites, the literature is quite scarce of examples and approaches, which are useful for practitioners.

A review of the methods for the selection of target sites for translocation emphasizes that microsite conditions play the important role in the success of translocation [11]. However, the selection approach that considers ecological features of a site (or more sites) at different scale should be preferred. Furthermore, it is important to consider the scale at which the selection is operating both spatially and temporally. For instance, the identification of target sites in the dry season may lead to the release of riparian plant species in areas that are long submerged during the wet season (Author's observation). For species with well-defined seasonal growth patterns, it is important that all ecological variables are considered in the context of the phenological phases. For instance, the reintroduction attempt of *Stratiotes aloides* L. failed because the water depth was too high (Author's observation). *S. aloides* floats in summer and sinks in autumn to overwinter as a rosette with green leaves on the sediment surface [29]. So, water depth is not very important during summer, but affects the plant lift to the water surface in spring. In some other cases, it was important to assure the right water level during critical phases of the growth cycle, for example to increase the flowering synchrony in emergent plants (plants living submerged but flowering above water), like *Ranunculus penicillatus* (Dumort.) Bab. Problems related to water level can be solved with the use of floating islands (see above).

Another issue related to a receptor site is the control of natural wetland dynamics. Receptor sites may require constant management for different reasons (e.g. slow down the natural vegetation dynamics, avoid landfill of the water body, reshaping of the river banks, etc.). Such efforts are expensive in a long-term perspective, so the selection of the target site should take into account this aspect in order to minimize the extra management.

Below we present some approaches that can be used to select suitable receptor sites at different spatial scales for aquatic species *s.s.* or riparian species.

The "current/historical growing site comparison" is the common method to assess the threats and to identify the target sites for the species. This approach compares the ecological characteristics of current sites with those of sites from where the species has disappeared. This method, if applied at a large spatial scale, has the advantage to provide important information on the species ecological niche, which may be useful to find out suitable growing sites. Moreover, the comparison of current and former sites allows the identification of environmental changes that brought the species to extinction. For instance, it highlights differences in water or soil quality (e.g. nutrient enrichment, accumulation of pollutant, etc.) between the current and extinct sites. This is an extensive sampling task, which may provide information on critical thresholds of certain parameters for the species. Above approach may provide enough data on critical environmental changes for the species during one sampling season, while usually long-term monitoring is required to find out such changes. However, as more parameters are included in the analysis, the higher number of sites is required to suffice data analysis. This can complicate the application of this method on very rare species that only have few sites. Moreover, in-depth analysis of several samples of water and/or soil may be expensive.

Another interesting approach is the use of logistic models based on few easy-to-measure parameters, which may yield preliminary information on the suitability of the several potential receptor sites, with a minor effort. As an example of this approach is the model for site selection developed by [30] for the choice of suitable reintroduction sites of the endemic quillwort *Isoëtes malinverniana* Ces. & De Not. Three variables (water electrical conductivity, water pH and water clarity) were used as predictors of the presence or absence of *I. malinverniana*. The model was based on the "current/historical growing site comparison" approach but with very few variables. This simplified model was particularly useful for a very rare species with a low number of extant populations, which precluded the application of statistical models based on a high number of variables. The selected variables were chosen among those of the highest ecological importance in predicting the suitability/unsuitability of potential reintroduction sites for the target species. Predictive variables differ among species, but the most important concept is that models based on few ecologically important parameters may represent cost-effective procedures for the selection of the receptor sites.

In addition, the water and soil characteristics determining the occurrence of the species may vary at the microsite level, affecting the distribution pattern of the target species (especially for riparian species). The role of microsite properties in the selection of the target site is crucial and the site that from the macroecological point of view may be suitable, may only partially fulfill the species requirements, or may consists of a mosaic of suitable and unsuitable land patches. One aim is to maximize the survival of the released plants, which can be obtained not only selecting the suitable sites among several potential sites, but also selecting suitable area within the suitable site. In the reintroduction attempt of *Aldrovanda vesiculosa* L. in Czech Republic, [31] selected suitable microsites on the basis of few variables like water depth and fluctuations, vegetation composition, the degree of shading, water transparency and water CO_2 concentration. The investigation of the microsite conditions of about 30 stands of *Leucojum aestivum* L. subsp. *aestivum* in North Italy aimed at reinforce some populations, revealed that within populations, the species occurs in niches with significantly higher water content (48.1% vs. 44.7%), lower content of sand (17.6% vs. 22.3%) and higher content of silt (59.7% vs. 55.6%) and nitrogen (0.55% vs. 0.49%) than microsites without the species. Interestingly, the above mentioned differences are quite weak, but enough to affect the spatial distribution of the species [32].

The next approach consists of simulation of conditions that may be favourable or unfavourable for the target species. For instance, in the reintroduction of *Marsilea quadrifolia* L. in agricultural landscape, the species tolerance to different types of herbicides was experimentally evaluated [33]. Although the experimental approach may provide valuable information on a certain aspect, it cannot replace the field approach.

CONSTRAINTS

Problems Due to Salt Intrusion

A problem that has been rising in many lowland wetlands is the seepage of salt water into freshwater bodies due to the rising of the sea level or the land subduction. This may change freshwater wetlands in brackish water areas, with very negative and potentially lethal consequences for the freshwater flora. This mechanism is frequent in deep quarries along the coast. A similar effect is mediated by periodical tidal surges that may affect coastal lagoons, estuarine environment and retrodunal ponds. The seepage of salt water is predicted to increase with increasing sea level (mediated by climate change), a phenomenon that will have long-term effect on translocation success [34].

Problems of Alien Species and of Native Predators

The interconnection of water bodies may certainly have positive effect on translocated populations, but may also represent a negative aspect, by favoring the spread of alien (sometimes invasive) species. This is one of the major problems to face during the different phases of a translocation, because detection (planning phase), eradication (site preparation) and management (post-translocation activity) of alien species may be an unequal battle. The alien species with the greatest impact on translocated aquatic and riparian plants in Europe are nutria (*Myocastor coypus* Molina) grass carp (*Ctenopharyngodon idella* Valenciennes), Louisiana crayfish (*Procambarus clarkii* Girard), and yellow-bellied slider (*Trachemys scripta* ssp. *elegans* e *Trachemys scripta* ssp. *scripta*).

Figure 8. Cages are used to exclude predators – Photo T. Abeli.

Figure 9. *Eichhornia crassipes* native in South America is invasive in Nepalese wetlands – Photo S. Orsenigo.

For instance, grass carp, native of eastern Asia has been introduced in Europe and North-America as a control for aquatic weeds, so its impact on aquatic plants is remarkable.

This problem applies also to native predators, if the populations are out of control. In particular, native fishes and waterfowls may be a threat for oligotrophic plants because they may enhance water nutrient enrichment, through their faeces. Waterfowls and aquatic mammals (e.g. beaver, musk rats, etc.) may also actively feed on translocated plants and uproot them. The control of alien species may be done with traps or active hunting. Additionally, the exclusion of them from the translocation areas is effective method (Figure 8), but some species may be particularly difficult to eradicate or control, like crayfishes.

Among the threats for wild and translocated wetland plant populations, there are several alien wetland plants that may affect both on riparian and strictly aquatic species. In Europe, the eradication of *Amorpha fruticosa* L. is particularly difficult, because it forms dense stands and can easily penetrate the native vegetation, also very dense reed communities. In water environment, *Ludwigia peploides* subsp. *montevidensis* (Spreng.) P.H. Raven, *Lagarosiphon major* Ridl. Moss ex Wager, *Elodea* spp. and *Nelumbo nucifera* Gaertn. are locally invasive. In contrast, *S. aloides* a threatened species in its native range in Western Europe, is invasive in North America and New Zealand. *Eichhornia crassipes* is strongly invasive in tropical and sub-tropical countries (Figure 9).

Pollution

Receptor sites located in agricultural or industrial contexts may be characterized by high level of pollutants or nutrients, and specific investigation of the tolerance of the target species to these substances may be required. Pollutants found in agricultural and industrial areas include herbicides, pesticides, oil derivatives, heavy metals, etc. Tolerance to high level of nutrients is also important for the target species released in wetlands close to crops.

Interestingly, some plant species (e.g. *Ceratophyllum demersum* L., *Elodea canadensis* Michx. and *Lemna minor* L.) can effectively remediate herbicide-contaminated waters [35], so they may be used to reduce the concentration of pollutant in the water body of the target site. However, the use of species for remediation should be done avoiding the release of alien or invasive species.

MONITORING WETLAND PLANT TRANSLOCATION

Reasons for Monitoring

The translocation site should never be abandoned afterwards, but it should be monitored for a number of years. The post-translocation monitoring can show problems in the plant establishment and may allow the last minute modifications to be done.

Moreover, the monitoring helps to judge whether the translocation has been successful or not. So there are two types of monitoring: the short-term monitoring aimed at correct contingent problems that may rise just after the translocation and the medium- to long-term monitoring aimed at evaluating the success of the translocation.

The first monitoring phase may last few months to one year with high frequency observations to possible corrections. In particular, for riparian species it is important to verify the soil moisture and if required, provide additional watering. For aquatic plants some water variables (e.g. pH, temperature, level, etc.) should be constantly monitored. Another important task is the monitoring of predators or alien invasive species not detected during the selection of the receptor site. It is also possible that species considered harmless may form unpredicted threats for the target species.

The long-term monitoring or the monitoring of success should be targeted to evaluate how viable is the translocated population by assessing mortality and growth rates of the population, the reproductive performance, the changes in vegetation cover and the capability of the population to expand into nearby habitats (Figure 10).

These are just examples and important parameters may differ among species. The monitoring of success usually does not require high frequency of sampling (e.g. once a year is enough). As said above, aquatic and riparian species are highly sensitive to environmental changes related to water physical and chemical parameters. Consequently, these parameters should be monitored in the long-term period, considering also external perturbations in the target site.

Figure 10. Monitoring the population of *Leucojum aestivum* three years after the translocation – Photo P. Cauzzi.

Methods of Monitoring

Methods of monitoring may differ depending on the initial scopes of the translocation. Moreover, monitoring must be based on collected data to answer one or a series of specific questions. Just "has it worked?" is not enough. It is important to understand how and why it has worked (e.g. does the species reproduce? Do the meta-population dynamics work? Is there gene flow between sub-populations? Is the population attractive for pollinators?). Other questions may depend on the target species. (e.g. in *Ranunculus tripartitus* DC. there is the need to demonstrate the seed-set in population within 40-year cycle, in *Populus nigra* L./*Alisma gramineum* Lej. there is the need to demonstrate seed germination).

The most common method to answer questions is the monitoring of population characteristics, in terms of growth and reproduction based on individual plant performance. Obviously, the techniques differ between strictly aquatic plants and riparian species. Monitoring of riparian species does not present particular challenges and common techniques available for terrestrial plant may be applied. To follow individual plant performance plants should be individually identifiable. There are several ways to identify individual plants within a population: 1) mark each plant with labels; 2) use the coordinates (for translocations in wide areas); 3) dispose the plants within the cells of a hypothetical grid or dispose the plants at regular distances along the transect [36, 37].

For strictly aquatic species above methods may not work, because plants may simply not be on hand (e.g. water depth, flow, etc.). Luckily, different monitoring methods are available for aquatic plants [38], for example direct observation in deep water by snorkeling or diving. For species with floating leaves like *Nymphaea*, *Nuphar,* etc. counting the number of plants should be done early in the growing season, when leaves are small and do not form extensive leaf carpets, and it is possible to distinguish single emergent individuals. An interesting method to monitor submerged plants in transparent waters is the use of floating grid that works as the grids used for terrestrial species.

Acknowledgments

Authors are grateful to Dr. Sandrine Godefroid (National Botanic Garden of Belgium, Bruxelles), Dr. Joyce Maschinski (Fairchild Tropical Botanic

Garden, Miami, Florida, USA) and Dr. Priptal S. Soorae (IUCN/SSC Reintroduction Specialist Group, Abu Dhabi) for information on translocation of aquatic plant species around the World. A special thank goes to Dr. Richard Lansdown (Ardeola Environmental Services, UK; IUCN/SSC Freshwater Plant Specialist Group, UK) for his precious suggestions on an early version of the manuscript and to Anne Jäkäläniemi (Oulanka Research Station, Thule Institute, University of Oulu, Finland) for her suggestions and linguistic revision.

REFERENCES

[1] Carpenter, S. R. & Lodge, D. M. (1986). Effects of submersed macrophytes on ecosystem processes. *Aquatic Botany*, 26, 341–370.

[2] Rantala, M. J., Ilmonen, J., Koskimäki, J., Suhonen, J. & Tynkkynen, K. (2004). The macrophyte, Stratiotes aloides, protects larvae of *dragonfly Aeshna viridis against* fish predation. *Aquatic Ecology,* 38, 77-82.

[3] Beintema, A. J., van der Winden, J., Baarspul, T., de Krijger, J. P. van Oers, K. & Keller K. (2010). Black terns Chlidonias niger and their dietary problems in Dutch wetlands. *Ardea,* 98, 365-372.

[4] Engelhardt, K. A. M. & Ritchie, M. E. (2001). Effects of macrophyte species richness on wetland ecosystem functioning and services. *Nature,* 411, 687-689.

[5] Bilz, M., Kell, S. P., Maxted, N. & Lansdown R. V. European Red List of Vascular Plants. Luxembourg, Publications Office of the European Union; 2011.

[6] IUCN (2010). Mediterranean Red List/Aquatic Plants. Available from URL: http://www.iucnredlist.org/initiatives/mediterranean/mediterraneanaquaticplants. Accessed: 25/03/2014.

[7] Rossi, G., Montagnani, C., Gargano, D., Peruzzi, L., Abeli, T., Ravera, S., Cogoni, A., Fenu, G., Magrini, S., Gennai, M., Foggi, B., Wagensommer, R. P., Venturella, G., Blasi, C., Raimondo, F.M. & Orsenigo, S. Lista Rossa della Flora Italiana. 1. Policy Species e altre specie minacciate. Roma: Comitato Italiano IUCN e Ministero dell'Ambiente e della Tutela del Territorio e del Mare; 2013a.

[8] Ricciardi, A. & Simberloff, D. (2009). Assisted colonization is not a viable conservation strategy. *Trends in Ecology and Evolution,* 24, 248-253.

[9] Godefroid, S., Piazza, C., Rossi, G., Buord, S., Stevens, A.-D., Aguraiuja, R., et al. (2011). How successful are plant species reintroductions? *Biological Conservation,* 144, 672–682.

[10] IUCN/SSC. Guidelines for Reintroductions and Other Conservation Translocations. Version 1.0. Gland, Switzerland: IUCN Species Survival Commission; 2013.

[11] Maschinski, J. & Haskins, K. E. Plant reintroduction in a changing climate, promises and perils. Washington DC: Island Press; 2012.

[12] Rossi, G., Amosso, C., Orsenigo, S. & Abeli, T. (2013b). Linee guida per la traslocazione di specie vegetali spontanee. Quad. Cons. Natura, 28, Roma, MATTM – Ist. Sup. Protezione e Ricerca Ambientale (ISPRA).

[13] Soorae, P. S. editor (2013). Global Re-introduction Perspectives: 2013. Further case studies around the globe. Gland, Switzerland: IUCN/SSC Re-introduction Specialist Group and Abu Dhabi, UAE: Environment Agency-Abu Dhabi. Xiv + 282 pp.

[14] Rout, T. M., Hauser, C. E. & Possingham, P. (2009). Optimal adaptive management for translocation of a threatened species. *Ecological Application, 19*, 515–526.

[15] Gornall, R. J., Hollingsworth, P. M. & Preston, C. D. (1998). Evidence for spatial structure and directional gene flow in a population of an aquatic plant, Potamogeton coloratus. *Heredity,* 80, 414–421.

[16] Russell, J. R., Weber, J. C., Booth, A., Powell, W., Sotelo-Montes, C. & Dawson, I. K. (1999). Genetic variation of Calycophyllum spruceanum in the Peruvian Amazon Basin, revealed by amplified fragment length polymorphism (AFLP) analysis. *Molecular Ecology,* 8, 199–204.

[17] Menges, E. S. (2008). Restoration demography and genetics of plants: when is a translocation successful? *Australian Journal of Botany,* 56, 187-196.

[18] Baskin, C. C. & Baskin, J. M. Seeds: ecology, biogeography and evolution of dormancy and germination. London, Academic Press; 1998.

[19] Abeli, T., Parolo, G., Mondoni, A. & Rossi, G. (2011). Bridging the gap between seed banks and in situ conservation. *Samara,* 20, 6.

[20] Magrini, S., Olmati, C., Onofri, S. & Scoppola, A. (2010). Recovery of viable germplasm from herbarium specimens of Osmunda regalis L. *American Fern Journal,* 100, 159–166.

[21] Cochrane, J. A., Crawford, A. D. & Monks, L. T. (2007). The significance of ex situ seed conservation to reintroduction of threatened plants. *Australian Journal of Botany,* 55, 356–361.

[22] Vallee, L., Hogbin, T., Monks, L., Makison, B., Matthes, S. M. & Rossetto, M. Guidelines for the traslocation of threatened plants in Australia. Second edition. Canberra: Australian Network for Plant Conservation; 2004.

[23] Smolders, A. J. P., Lamers, L. P. M., den Hartog, C. & Roelofs, J. G. M. (2003). Mechanisms involved in the decline of Stratiotes aloides L. in the Netherlands: sulphate as a key variable. *Hydrobiologia,* 506/509, 603-610.

[24] Abeli, T., Orsenigo, S., Ardenghi, N. M. G., Lucassen, E. C. H. E. T & Smolders, A. J. P. (2014). Hydrochemical characterization of a stand of the threatened endemic Isoëtes malinverniana. *American Fern Journal,* 103, 241-244.

[25] Guerrant, E. O. & Kaye, T. N. (2007). Reintroduction of rare and endangered plants: common factors, questions and approaches. *Australian Journal of Botany,* 55, 362–370.

[26] Muzzi, E. & Rossi, G. Il recupero e la riqualificazione ambientale delle cave in Emilia-Romagna. Manuale tecnico-pratico. Bologna: Regione Emilia-Romagna; 2003.

[27] Prashant, D. R., Billore, S. K., Sharma, J. K., Singh, N. & Ram, H. (2013). Treatment of wastewater and restoration of aquatic systems through an eco-technology based constructed treatment wetlands. A successful experience in Central India. *Water Science and Technology,* 68, 1566-1573.

[28] Nakamura, K. and Mueller, G. (2008). Review of the Performance of the Artificial Floating Island as a Restoration Tool for Aquatic Environments. World Environmental and Water Resources Congress 2008: pp. 1-10.

[29] Roelofs, J. G. M. (1991). Inlet of alkaline river water into peaty lowlands: effects on water quality and Stratiotes aloides L. stands. *Aquatic Botany,* 39, 267-293.

[30] Abeli, T., Barni, E., Siniscalco, C., Amosso, C. & Rossi, G. (2012). A cost-effective model for quick site evaluation for the reintroduction of a threatened quillwort. Aquatic Conservation: *Marine and Freshwater Ecosystems,* 22, 66-73.

[31] Adamec, L. & Lev, J. (1999). The introduction of the aquatic carnivorous plant aldrovanda vesiculosa to new potential sites in the czech republic: a five-year investigation. *Folia Geobotanica,* 34, 299-305.

[32] Parolo, G., Abeli, T., Rossi, G., Dowgiallo, G., Matthies, D. (2011). Biological Flora of Central Europe: Leucojum aestivum L. Perspectives in Plant Ecology, *Evolution and Systematics,* 13, 319-330.

[33] Bruni, I., Gentili, R., De Mattia, F., Cortis, P., Rossi, G. & Labra, M. (2013). A multi-level analysis to evaluate the extinction risk and conservation strategy for the aquatic fern Marsilea quadrifolia L. in Europe. *Aquatic Botany,* 111, 35-42.

[34] Eliot, I., Finlayson, C. M. & Waterman, P. (1999). Predicted climate change, sea-level rise and wetland management in the Australian wet-dry tropics. *Wetlands Ecology and Management*, 7, 63-81.

[35] Rice, P. J., Anderson, T. A. & Coats, J. R. Phytoremediation of soil and water contaminants. In: Rice, P.J., Anderson, T.A. & Coats, J.R. editors. Phytoremediation of herbicide-contaminated surface water with aquatic plants. Washington DC: ACS Publications; 1997; 133-151.

[36] Smith, T. E. (2003). Observations on the experimental planting of Lindera melissifolia (Walter) Blume in Southeastern Missouri after ten years. *Castanea,* 68, 75-80.

[37] Elzinga, C. L., Salzer, D. W., Willoughby, J. W. & Gibbs, J. P. Monitoring plant and animal populations. Oxford, Blackwell; 2001.

[38] Madsen, J. D. & Wersal, R. M. A review of aquatic plant monitoring and assessment methods. Aquatic Ecosystem Restoration Foundation; 2012.

In: Macrophytes
Editor: Raven Capello
ISBN: 978-1-63321-518-4

Chapter 2

RESPONSES OF FLOATING MACROPHYTES TO DIFFERENT ENVIRONMENTAL STRESSORS

Cinzia Forni*

Department of Biology, University of Rome "Tor Vergata", Via della Ricerca Scientifica, Rome, Italy

ABSTRACT

During their life plants have to cope with natural or anthropogenic environmental stressors, including different pollutants. The exposure of aquatic macrophytes to these compounds mobilizes or generates, directly or indirectly, diverse signalling molecules and regulates many processes that amplify and specify the physiological response through transcriptional and metabolic changes. Due to consistent enhancement of water pollution and in order to better preserve aquatic ecosystems, the knowledge of response mechanisms used by the plants to cope with stress is of increasing significance. In this paper, the stressor driven changes and responses induced by different pollutants and phytoremediation potential of some floating macrophyte species will be considered and discussed.

Keywords: Floating macrophytes, pollutants, stress

---

* Corresponding author: E-mail: forni@uniroma2.it.

1. Water Resources and Pollution

Reducing biodiversity loss, ensuring environmental sustainability and decreasing the proportion of people without access to safe drinking water are the Goals of the fundamental Millennium Development, established by the United Nations General Assembly in its Millennium meeting in 2000. Under this perspective, one of the major concerns is about the preservation of aquatic resources for both ecosystem integrity and human health. Water resources are directly and indirectly affected by anthropogenic activities (e.g. land-use misuses and spreading of point and non-point pollution) and natural factors (e.g. climate change).

Several compounds can be ascribed to the group of hazardous contaminants affecting water quality, and their amount in the ecosystems is increased. Even though many natural streams have historically received metal contaminations from anthropogenic sources (e.g. treated sewage discharges, mining operations and runoffs from metal-refining industries or urban storm water), in recent years water pollution with toxic metals has accelerated dramatically (Rai, 2012).

Surfactants can be also included in the list of important water contaminants; these molecules are designed to have cleaning or solubilisation properties with a wide range of domestic and industrial applications (Ying, 2006). They are a diverse class of chemicals, that generally consist of a polar head group (either charged or uncharged), which is well solvated in water, and a non polar hydrocarbon tail, which is not easily dissolved in water (Ying, 2006). Problems can arise when these molecules are present in the aquatic ecosystems in relatively high concentration. Surfactants and their products are mainly discharged into sewage treatment plants and then dispersed through effluent discharge into surface waters and sludge disposal on soils. Their different behavior and fate in the environment are due to the type of molecules, i.e. nonionic and cationic surfactants have much higher sorption on soil and sediment than anionic ones (Ying, 2006 and references therein). Although in the environment most surfactants can be degraded by microbes, some of them may be persistent under anaerobic conditions, representing an important hazard that affects the aquatic organisms at different levels. Their harmful effects are depending on the type and concentration and become usually evident at levels in excess of environmental concentrations (Ying, 2006). Even though there is an awareness about the obvious implications for the fate and transport of these molecules in aquatic systems, the main body of literature has recently focused on their potential use to remediate contaminated environments for their ability to modify the behavior of other, preexistent or co-disposed contaminants (Hari et al., 2005; Jones-Hughes and Turner, 2005; Karapanagioti et al., 2005; Yang et al., 2002; Zhang and Zhou, 2005). For example, the sorption of cationic surfactants to a variety of synthetic and natural solids is an effective means of immobilising neutral and ionisable organic chemicals in landfill leachate as well as contaminated waters; whereas the application of either non-ionic or ionic surfactants, in concentrations above their critical micelle concentrations (CMCs), has been tested to release neutral chemicals and to enhance heavy metals removal from aquifers and soils (Turner and Xu, 2008). Therefore, the toxic effects exhibit by some surfactants, albeit at levels in excess of environmental concentrations, lead to consider very carefully their use in remediation.

Pharmaceutically active compounds (PhACs) are another important emerging water contaminants that are introduced into the environment via various human activities, including direct disposal of unused medication, or release from pharmaceutical manufacturing plants, hospitals and veterinary drugs. Most of concern is on antibiomicrobial drugs, one of the most important groups of PhACs that have been used in large quantities for some decades. Besides hospitals, municipal sewage and sewage water treatment plants, other important source of antimicrobial contamination is intensive farming, including aquaculture, where they are used both as feed additive and in mass therapy (Forni et al., 2006). Animal waste contains quinolones and other drugs in active form, because of the slow absorption rates and excretion with deject (Cascone et al., 2004 and references therein). Furthermore, uneaten medicated feed contribute to environmental contamination: it has been demonstrated that the major part of drugs administered as medicated feed during a treatment period, ends up in the environment (Cascone et al., 2004 and references therein). In addition, agricultural practice, i.e. normal practices of animal sludge application on soil can lead to the introduction of these chemicals into the terrestrial (soil top dressing) and aquatic (run off of top dressed soils) compartments (Jørgensen and Halling-Sørensen, 2000). Within the last decade, an increasing number of studies covering antibiotic input, occurrence, fate and effects have been published (Heberer, 2002; Kümmerer, 2009 and references therein). Due to the fact that some compounds are not efficiently removed during wastewater treatment processes, being able to reach surface, groundwater and drinking waters, their occurrence in these waters has been reported (Delépée et al., 2004 and references therein; Onesios et al., 2009; Verlicchi and Zambello, 2014 and references therein). In fact, the conventional wastewater treatment plants (WWTPs) can effectively reduce most of the molecules present, e.g. carbon and nitrogen content of raw sewage, but the removal of PhACs residues is often insufficient and WWTP effluents are frequently pointed out as the main source of these microcontaminants.

2. Biological Effects of Pollutants on Aquatic Macrophytes

The control of pollution sources is needed to reduce the environmental impact of contaminants, and in the meantime the development of an effective surface water monitoring network will be a critical element for any protection or restoration project. Pollution can be measured by physical and chemical methods giving values which are verified to comply with regulation requirements (APHA, 1998). Nevertheless, these methods present some limitations like, for example, the quantification of compounds present in low amounts, i.e. below analytical detection limits. Therefore, the use of living organisms that accumulate pollutants overcomes some of these difficulties and allows the characterization of contaminant bioavailability (Mouvet et al., 1993; Dazy et al., 2009).

In aquatic ecosystems macrophytes share the position of primary producers in the food chain and they play an important role in the structural and functional aspects of these ecosystems by changing water movement regimes, providing shelter to other components of the ecosystem, and altering water quality, for example by regulating oxygen balance and nutrient cycles (Sood et al., 2012). Moreover, macrophytes are among the first not target organisms reached by pollutants in these environments. Such plants are used as *in situ* biomonitors of water pollution because of their abundance and limited mobility (Tel-Or and Forni, 2011 and references therein). Some species belonging to the genus *Lemna* are also used as model plants in ecotoxicological studies (Tel-Or and Forni, 2011 and references therein).

a. Macrophytes vs. Heavy Metals

Heavy metals (HM) are environmental pollutants released from both industrial and agricultural sources affecting the biosphere in many places worldwide. Aquatic ecosystems are more sensitive to pollutants than terrestrial ones because most uptakes are facilitated by dissolution in water, which may lead to an accumulation of HM (Parlak and Yilmaz, 2013).

HM contamination may pose a serious threat to these ecosystems and human health. Alkaline precipitation, ion exchange columns, electrochemical removal, filtration, and membrane technologies are the currently available technologies for HM removal; but these conventional methods are not economical and may produce adverse impacts on aquatic environment.

The plants are capable to remove many toxic substances from reservoirs, acquiring the pollutants from waters as food elements, basically through root system using the products of decomposition for their viability (Watai et al. 2004; Tel-Or and Forni, 2011). The bioaccumulation characteristic has been exploited in phytoremediation technology and became an effective method of environment clearing due to the plants ability to take up the contaminants, sometimes at the concentration level thousands times higher than background one. Phytoremediation of metals is a cost-effective "green" technology based on the use of specially selected metal-accumulating plants to remove toxic metals from soils and water (Tel-Or and Forni, 2011 and references therein). Some of these HM, such as As, Cd, Hg, Pb or Se, are not essential, since they do not perform any known physiological function in plants. Others, such as Co, Cu, Fe, Mn, Mo, Ni and Zn, are essential elements required for normal growth and metabolism of plants. Water plants actively accumulate Pb, Cd, Cu, Fe, Mn, Zn, Cr, and other elements and substances; *Typha*, *Phragmites*, *Eichhornia*, *Azolla*, *Lemna*, and other aquatic macrophytes represent potent wetland plants used for HM removal (Rai, 2008; Tel-Or and Forni, 2011 and references therein). Even though the disposed biomass of macrophytes may be used for various fruitful applications (Rai, 2008), biomass disposal problem and seasonal growth of plants are some of the limitations in the successful transfer of phytoremediation technology from the laboratory to the field.

In contemporary literature many studies are focused on the actual metals uptake by plants but sufficiently less is known about the process of their transport and transformation inside the plant's tissue. HM phytotoxicity may induce alterations of numerous physiological processes at cellular/molecular level by inactivating enzymes, blocking functional groups of metabolically important molecules, displacing or substituting for essential elements and disrupting membrane integrity. Plants, that have to cope with this toxicity, mobilize or generate several signalling molecules and regulate many processes that involve physiological response through transcriptional and metabolic changes (Rascio and Navari-Izzo, 2011 and references therein). Defence mechanisms give rise to series of responses to control uptake, accumulation and translocation of these dangerous elements and detoxify them by excluding the free ionic forms from the cytoplasm. The ability of the plants to accumulate metals is apparently due to their association with the cell wall which limits their penetration into the body; at the initial stage the metals are absorbed by the roots and then progressively redistributed within the plant. Plants try to hinder the entrance of heavy metals into root cells and one of the strategies lies in the entrapment of metals in the apoplastic environment as dynamic physical barrier (Ovečka and Takáč, 2014) or to anionic groups of cell walls. Studies undertaken on floating macrophytes elucidated the role of the cell wall components in the HM binding. In the water fern *Azolla* pectin plays a major role in the binding of Pb^{+2} and Sr^{+2} to the cell wall, in a process that exchanges Ca^{+2} and Mg^{+2} for Pb^{+2} and Sr^{+2} (Cohen-Shoel et al. , 2002; Benaroya et al. , 2004; Schor-Fumbarov et al., 2005; Tel-Or and Forni, 2011 and references therein). This is based on the ion exchange capability of the fern and demonstrates the higher affinity of the toxic metals (i.e. Pb^{+2}, Sr^{+2} and others) to the pectin matrix as compared to not toxic ones, like Ca^{+2} and Mg^{+2}. Besides pectins, phosphoinositol in phytic acid exhibits high affinity to toxic ions in the plant matrix. The biosynthesis of phoshoinositol by L-Myo-Inositol-phosphate was demonstrated by Benaroya et al. (2004a). While, Rakhshaee et al. (2009) used potentiometric titration as analytical method to demonstrate the role of the ratio –COOH/–COO– on the heavy metals uptake, and they found an increase of the removal of HM ions by increasing the carboxylate from 0.92 to 2.42 mmol g^{-1} *Lemna minor* L.

Whenever the plants are not able to avoid metal uptake, most of the HM, that do enter the plant, are then kept in root cells, where they are detoxified by complexation with several molecules as amino acids, organic acids or metal-binding peptides and/or sequestered into vacuoles. The role of metal binding proteins has been demonstrated in *Azolla* by Schor-Fumbarov et al. (2005), who found that the metallothionein may function in metal ion mobilisation in the *Azolla* tissues; Schneider-Nahum (2000) also suggested the involvement of phytochelatins or similar short peptides in Cd^{+2} binding by the fern.

The presence of HM causes a stress to plants that evokes parallel and/or consecutive events, in which every mechanism could be at the same time cause and effect of metabolic changes. Stunted growth, chlorosis and necrosis are some of the visible symptoms indicating severe metal phytotoxicity, and dose-response curves can depict the effect of heavy metals on plants (Hagemeyer, 2004). Duckweeds (*Lemna* spp.), which often forms dense floating plants in aquatic ecosystems, such as ditches, ponds and lakes, are often selected for study and can be used as indicator species to assess ecotoxicity of waters polluted by contaminants. From a visual assessment of affected plants, it is clear that change in plant pigments is in different ways depending on the mode of action of the toxicant. Biomarkers such as pigment content, primary chlorophyll and carotenoids, are commonly used as endpoints for toxicity tests. Chlorophyll content is one of the recommended endpoints in the standard of the International Organisation for Standardisation (ISO) and the Organisation for Economic Co-operation and Development (OECD) guideline for *Lemna*-tests (OECD, 2002; Naumann et al., 2007) and is frequently used as an endpoint for other macrophyte species (Cedergreen et al., 2007). An example of the effect of HM on chlorophylls content in *L. minor* after 7 days of treatment with different concentrations of lead and cadmium is shown in Figure 1.

The phytotoxicity of metals arises partly from the generation of reactive oxygen species (ROS), which can heavily damage cellular metabolism. In fact, the ROS , i.e. $O_2^{\cdot -}$, H_2O_2, $^1O_2$, $HO_2^{\cdot -}$, $OH^{\cdot}$, ROOH, $ROO^{\cdot}$, and $RO^{\cdot}$ affect many cellular functions by damaging nucleic acids, oxidizing proteins, and causing lipid peroxidation (LPO) (Gill and Tuteja, 2010 and references therein). Whether ROS will act as damaging, protective or signaling factors depends on the delicate equilibrium between their production and scavenging. The equilibrium between the production and the scavenging of ROS may be perturbed by various HM. Stress-induced ROS accumulation is counteracted by enzymatic antioxidant systems that include a variety of scavengers, such as the antioxidant enzymes and non-enzymatic low molecular metabolites, like phenolics (Dazy et al., 2009; Gill and Tuteja, 2010 and references therein), that can be potential markers of non-visible plant damage (Pasqualini et al., 2003). Changes in phenolics content of water fern *Azolla filiculoides* Lam. and duckweed *Lemna minor* L. exposed to different concentration of lead and cadmium are reported in Tables 1 and 2; *Azolla* grown in Pb medium increased phenolics production in concentration-dependent manner, reaching a maximum (9.97 ± 1.28 μg / mg.f.w.) at 7 days of exposure to 10 ppm (Table 1). *Lemna* plants treated with Pb or Cd showed an increase in phenolics respect to untreated ones (Table 2). Within time phenolics production was almost constant at all Pb concentrations tested. In plants exposed to Cd, a peak was detected at 2.5 ppm Cd, while phenolics content decreased significantly ($p<0.01$) at 10 ppm Cd at 7th day of treatment with values comparable to the control plants (Table 2).

Table 1. Effects of different concentrations of Pb and Cd on phenolic compounds production in *A. filiculoides*. Data are expressed in μg chlorogenic acid equivalent (μg /mg fresh weight) and are the means ± SE (n = 9)

HM	ppm	Phenolics content (μg /mg.f.w.)	
		0 days	7 days
Pb	0	1.79 ± 0.13	1.55 ± 0.14
	2.5	1.88 ± 0.16	3.99 ± 0.63
	5	1.77 ± 0.15	6.89 ± 0.99
	10	1.99 ± 0.15	9.97 ± 1.28
Cd	0	1.97 ± 0.10	1.70 ± 0.098
	2.5	1.78 ± 0.16	9.99 ± 1.44
	5	1.57 ± 0.18	9.88 ± 1.95
	10	2.00 ± 0.34	8.16 ± 1.33

Table 2. Effects of different concentrations of Pb and Cd on phenolic compounds production in *L. minor* . Data are expressed in μg chlorogenic acid equivalent (μg /mg fresh weight) and are the means ± SE (n = 9)

HM	ppm	Phenolics content (μg /mg.f.w.)	
		0 days	**7 days**
Pb	0	2.57 ± 0.94	2.35 ± 0.43
	2.5	2.46 ± 0.74	10.97 ± 2.05
	5	2.35 ± 0.26	12.13 ± 2.76
	10	2.37 ± 0.21	9.48 ± 0.92
Cd	0	2.57 ± 0.24	2.34 ± 0.32
	2.5	2.46 ± 0.35	12.83 ± 2.46
	5	2.35 ± 0.24	6.85 ± 0.85
	10	2.75 ± 0.27	3.63 ± 0.44

Non-enzymatic defence also involves glutathione, ascorbic acid, α-tocopherol, β-carotene and other compounds capable of quenching ROS, while the enzymatic defence include superoxide dismutases, catalases, peroxidases, as guaiacol peroxidase (G-POD) and ascorbate peroxidase (APX), glutathione reductase and NADP+ reducing enzymes (Mittler, 2002; Scandalios, 2002;

Razinger et al., 2006). Therefore, plant stress tolerance may be improved by the enhancement of levels of antioxidant enzymes as well as antioxidant metabolites. Enhancement of peroxidase and PPO activities has been reported in *Lemna* and *Nymphaea* exposed to heavy metals (Lavid et al., 2001; Cvjetko et al., 2010).

Most of the studies on macrophytes response to HM have considered the exposure to single heavy metal, and just few are reporting the effects of more than one. Various interactions can occur when the plants are exposed to more than one unfavorable element. Such interaction has been described as independent, additive, synergistic or antagonistic (Shaw et al., 2004 and references therein). In general, these effects do not fit the toxicological profiles of individual metals making a combination and depending on their concentrations and factors such as temperature, pH, and light (Wilde et al., 2006). The uptake rates may be modified respect to those detected with a single metal (Saygidğer and Doğan, 2004; Shaw et al., 2004; Cvjetko et al., 2010; Neagu et al., 2014) and this may be explained by the competition between the metals for the binding sites of plant cells. However, similarly to single metal, the presence of more than one HM affect the amount and the ratio of photosynthetic pigments (Figure 1) and enhances phenolics production (Figure 2) in *L. minor* plants exposed to cadmium and lead for seven days.

The interactions between copper (in the concentrations of 2.5 μmol l^{-1} and 5 μmol l^{-1}) and cadmium (5 μmol l^{-1}) in common duckweed (*Lemna minor*) have been investigated by exposing the plants to either metal or to their combinations for four or seven days (Cvjetko et al., 2010). By comparing single to combined HM treated plants changes in the responses were detected. According to the authors the complex response detected when both metals were present in the growth medium cannot be explained by simple antagonism and/or synergism.

Progress in understanding the way through which HM are taken up has been made, but there is still a great need in strengthening the knowledge concerning metal localization and binding inside the plant, which would help to manage the process of pollutants extraction from waters, especially when more than one HM is present in contaminated waters. To do this it is necessary to explore the dynamics of pollutant accumulation, distribution in tissues and organs and at last to identify the classes of compounds in which the metals are transported and deposited within the plant organs.

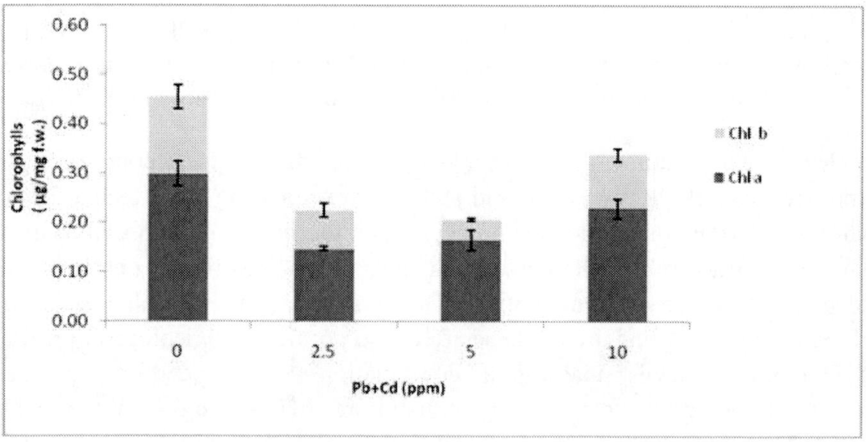

Figure 1. Effects of lead and cadmium on photosynthetic pigments of *Lemna minor* exposed for 7 days. Data are expressed in μg chlorophyll/mg fresh weight (f.w.) and are the means ± SE (n = 3).

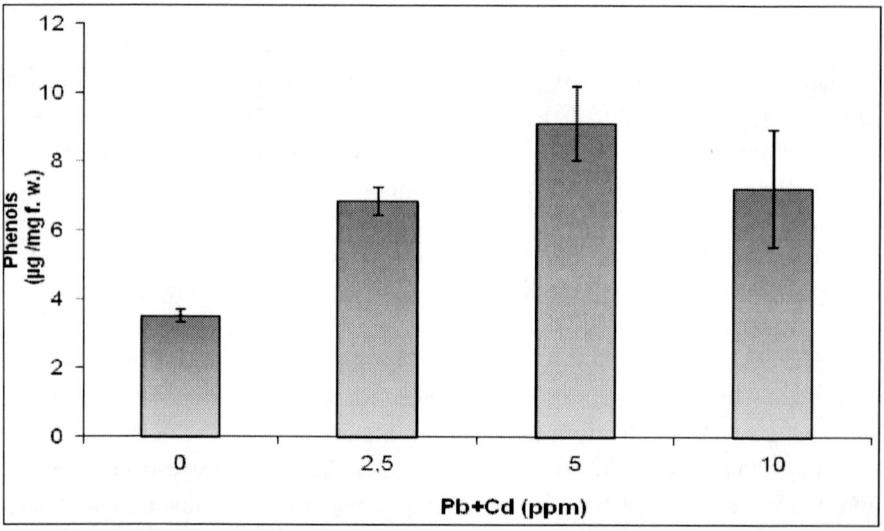

Figure 2. Effect of lead and cadmium on phenolics production in *L.minor* plants treated for seven days. Data are expressed in μg chlorogenic acid equivalent /mg fresh weight (μg /mg f.w.) and are the means ± SE (n = 3).

b. Macrophytes vs. Surfactants/Drugs

Surfactants and their degradation products enter the environment through the discharge of sewage effluents into surface waters and application of sewage sludge on land, and have the potential to impact the ecosystem owing to their toxicity on organisms. The toxicity data from laboratory and field studies are essential for us to assess the possible environmental risks from the surfactants (Ying, 2006). Biodegradation is an important process to treat surfactants in sewage treatment plants, and their degradation through microbial activity is the primary transformation occurring in the environment that can also enhances the removal of these surfactants, thus reducing their impact on biota. But many chemical and environmental factors may affect biodegradation of a surfactant, and the most influencing ones are the chemical structure, and physiochemical conditions of the environmental media. As a consequence the persistence of not or partially degraded surfactants in aquatic ecosystems causes adverse effects to the aquatic organisms (Tabor and Barber, 1996; Shang et al., 1999; Ying, 2006).

Literature data show that floating macrophytes, i.e. duckweed (*Lemna minor*) (Dilrigen and Ince, 1995; Forni et al. 2008) and water velvet (*Azolla filiculoides*) (Forni et al., 2008), exposed to the anionic surfactant sodium dodecyl sulphate (SDS), have a lower growth rate and chlorophyll contents, although both macrophytes were able to uptake the surfactant showing a remarkable effectiveness in phytoremediation (Forni et al., 2008). The stress response to SDS has been investigated by considering markers of abiotic stress, i.e. the activity of guaiacol peroxidase (G-POD), enzyme involved in oxidative stress response, as well as the enzymes of phenylpropanoid metabolic pathway phenylalanine ammonia-lyase (PAL), polyphenol-oxidase (PPO) (Forni et al., 2012). The presence of SDS within the duckweed affected enzymatic activities and phenol content. While in the fern phenolics amount, PAL, G-POD and PPO activities were not affected by SDS except for 100 ppm SDS, the only concentration that was taken up by the plants and not completely degraded.

Floating macrophytes can be an interesting tool also for *in situ* remediation of drug contaminated waters (Forni et al., 2006). Interesting results have been obtained in laboratory model studies with *Azolla filiculoides* exposed to an environmental persistent antibiotic commonly used in intensive farming, sulphadimethoxine (Forni et al., 2002). In a 5 week experiment, plants were cultivated outdoor at four drug concentrations (50, 150, 300 and 450 mg l^{-1}) in N-free mineral medium. Plant growth rate and morphology, N_2-

fixation and heterocyst frequency of the cyanobiont were affected by the drug, but the plants survived and actively removed sulphadimethoxine from the medium. Drug accumulation in the biomass was in order of magnitude up to mg g^{-1} plant dry weight (1000 ppm), moreover a positive relation between uptake and degradation rates with increased drug concentrations in the culture medium was detected. Basing on very high efficacy of the model, *Azolla* can be taken into consideration as a tool for sulphonamides environmental monitoring and decontamination (Forni et al., 2002). Ecotoxicological test was performed in order to detect the phytotoxicity on *L. minor* of flumequine. This quinolone antibiotic is widely used in intensive aquaculture both in marine and freshwater environments (Quevauviller and Maier, 1994), being also quite persistent in environmental compartments (approx. 150 days half-life, in surface sediment) (Hektoen et al., 1995). Flumequine, at all tested concentrations between 50 and 1000 $\mu g\ l^{-1}$, affected the plant growth and leaves and roots were damaged, but duckweed continued to grow on a five weeks period (Cascone et al., 2004). Moreover, increasing drug concentrations decreased the chlorophyll *b* content in plants. These effects depend on flumequine uptake by plants, which is quite high (from 0.72 to 13.93 $\mu g\ g^{-1}$ plant dry weight). Based on bioaccumulation activity, *Lemna* can be taken into consideration as a tool for *in situ* remediation of flumequine contaminated waters.

FUTURE DIRECTIONS

It is clear that, in spite of a growing public interest and the success of several pilot studies and field scale applications, more case studies should be undertaken to implement or to integrate the accumulated knowledge on the impact of contaminants on aquatic macrophytes. Moreover for a successful decontamination of heterogeneous pollution of waters, it is needed taking into consideration the chemistry of the specific polluted habitat and the complex interactions between contaminants, plant roots, and microorganisms. In some cases the mechanisms of tolerance, uptake of contaminants and metabolic pathway involved in the sequestration/degradation need to be clarified.

Further, more data are still needed to quantify the underlying economics, as a support for public acceptance and last but not least to convince policy makers and stakeholders (who are not very familiar with such techniques).

REFERENCES

APHA, (1998). Standard Methods for the Examination of Water and Wastewater. American Public Health Association, Washington, DC.

Benaroya, R., Tzin, V., Tel-Or, E. & Zamski, E. (2004). Lead accumulation in the aquatic fern Azolla filiculoides. *Plant Physiol. Biochem.*, *42*, 639-645.

Benaroya, R., Zamski, E. & Tel-Or, E. (2004a). L-Myo-inositol 1-phosphate synthase in the aquatic fern *Azolla filiculoides*. *Plant Physiol. Biochem.*, *42*, 97-102.

Cascone, A., Forni, C. & Migliore, L. (2004). Flumequine uptake and the aquatic duckweed, *Lemna minor* L.. *Water, Air Soil Poll.*, *156*, 241-249.

Cedergreen, N., Abbaspoora, M., Sørensenb, H. & Streibige, J. C. (2007). Is mixture toxicity measured on a biomarker indicative of what happens on a population level? A study with *Lemna minor*. *Ecotoxicol. Environ. Saf.*, *67*, 323-332.

Cohen-Shoel, N., Ilzycer, D., Gilath, I. & Tel-Or, E. (2002). The involvement of pectin in Sr^{+2} biosorption by *Azolla*. *Water Air Soil Poll.*, *135*, 195-200.

Cvjetko, P., Tolić S., Šikić S., Balen Biljana B., Tkalec M.,Vidaković- Cifrek Ž. & Pavlica, M. (2010). Effect of copper on the toxicity and genotoxicity of cadmium in duckweed (*Lemna minor* L.). *Arh. Hig. Rada Toksikol.*, *61*, 287-296

Dazy, M., Masfaraud, J. F. & Férard, J. F. (2009). Induction of oxidative stress biomarkers associated with heavy metal stress in *Fontinalis antipyretica* Hedw. *Chemosphere*, *75*, 297-302.

Delépée, R., Pouliquen, H. & Le Bris, H. (2004). The bryophyte *Fontinalis antipyretica* Hedw. bioaccumulates oxytetracycline, flumequine and oxolinic acid in the freshwater environment. *Science Total Environ.*, *322*, 243–253

Dilrigen, N. & Ince, N. (1995). Inhibition effect of anionic surfactant SDS on duckweed, *Lemna minor* with consideration of growth and accumulation. *Chemosphere*, *31*, 4185–4196.

Fendinger, N. J., Begley, W. M., Mcavoy, D. C. & Eckhoff, W. S. (1995). Measurement of alkyl ethoxylate surfactants in natural waters. *Environ. Sci. Technol.*, *29*, 856– 63.

Forni, C., Chen, J., Tancioni, L. & Grilli Caiola, M. (2001). Evaluation of the fern *Azolla* for growth, nitrogen and phosphorus removal from wastewater. *Water Res.*, *35*, 1592-1598.

Forni, C., Cascone, A., Fiori, M. & Migliore, L. (2002). Sulfadimethoxine and *Azolla filiculoides* Lam. A model for drug remediation. *Water Res.*, *36*, 3398-3403.

Forni, C., Patrizi, C. & Migliore, L. (2006). *Floating aquatic macrophytes as a decontamination tool for antimicrobial drugs*. In Twardowska, I., Allen, H. E., Haggblom, M. M. (Eds.), Soil and Water Pollution Monitoring, Protection and Remediation, NATO Science Series IV. *Earth and Environmental Sciences*, Vol. *69*, 467-477. Springer, The Netherlands.

Forni, C., Giordani, F., Pintore, M. & Campanella, L. (2008). Effects of Sodium Dodecyl Sulphate on the aquatic macrophytes *Azolla* and *Lemna*. *Plant Biosystems*, *142*/3, 665-668.

Forni, C., Braglia, R., Harren, F. J. M. & Cristescu, S. M. (2012). Stress responses of duckweed (*Lemna minor* L.) and water velvet (*Azolla filiculoides* Lam.) to anionic surfactant sodium-dodecyl-sulphate (SDS). *Aquatic Toxicol.*, 110-111, 107-113.

Gill, S. S. & Tuteja N. (2010). Reactive oxygen species and antioxidant machinery in abiotic stress tolerance in crop plants. *Plant Physiol. Biochem.*, *48*, 909-930.

Hagemeyer, J. (2004). *Ecophysiology of plant growth under heavy metal stress*. In: Prasad M.N.V. (Ed.), Heavy metal stress in Plants. From biomolecules to ecosystems, 2nd Edition. Springer, 201-222.

Hari, A. C., Paruchuri, R. A., Sabatini, D. A. & Kibbey, T. C. G. (2005). Effects of pH and cationic and non-ionic surfactants on the adsorption of pharmaceuticals to a natural aquifer material. *Environ. Sci. Technol.*, *39*, 2592–2598.

Heberer, T. (2002). Occurrence, fate, and removal of pharmaceutical residues in the aquatic environment: a review of recent research data. *Toxicol. Lett.*, *131*, 5-17.

Hektoen, H., Berge, J. A., Hormazabal, V. & Yndestad, M. (1995). Resistance of antibacterial agents in marine sediments. *Aquaculture*, *133*, 175–184.

Jones-Hughes, T. L. & Turner, A. (2005). Sorption of ionic surfactants to estuarine sediment and their influence on the sequestration of phenanthrene. *Environ. Sci. Technol.*, *39*, 1688–1697.

Jørgensen, S. E. & Halling-Sørensen, B. (2000). Drugs in the environment. *Chemosphere*, *40*, 691–9.

Karapanagioti, H. K., Sabatini, D. A. & Bowman, R. S. (2005). Partitioning of hydrophobic organic chemicals (HOC) into anionic and cationic surfactant-modified sorbents. *Water Res.*, *39*, 699–709.

Kümmerer, K. (2009). Antibiotics in the aquatic environment – A review – Part II. *Chemosphere*, *75*, 435-441.

Lavid, N., Schwartz, A., Yarden, O. & Tel-Or, E. (2001). The involvement of polyphenols and peroxidase activities in heavy-metal accumulation by epidermal glands of waterlily (Nymphaceae). *Planta*, *212*, 323-331.

Mittler, R. (2002). Oxidative stress, antioxidants and stress tolerance. *Trends Plant Sci.*, *7*, 405-410.

Mouvet, C., Morhain, E., Sutter, C. & Couturieux, N. (1993). Aquatic mosses for the detection and follow-up of accidental discharges in surface waters. *Water Air Soil Pollut.*, *66*, 333–348.

Naumann, B., Eberius, M. & Appenroth, K. J. (2007). Growth rate based dose–response relationships and EC-values of ten heavy metals using the duckweed growth inhibition test (ISO 20079) with *Lemna minor* L. clone St . *J. Plant Physiol.*, *164*, 1656-1664.

Neagu, D., Arduini, F., Quintana Calvo, J., Di Cori, P., Forni, C. & Moscone, D. (2014). Disposable electrochemical sensor to evaluate the phytoremediation of the aquatic plant *Lemna minor* L. towards Pb^{2+} and / or Cd^{2+}. Environ. Sci. and Technol.: in press. Dx.doi.org/10.1021/es500675x.

OECD, (2002). Guidelines for the testing of chemicals. *Lemna* sp. Growth Inhibition Test, Draft guideline 221.

Onesios, K. M., Yu, J. T. & Bouwer, E. J. (2009). Biodegradation and removal of pharmaceuticals and personal care products in treatment systems: a review. *Biodegradation*, *20*, 441-466.

Ovečka, M. & Takáč, T. (2014). Managing heavy metal toxicity stress in plants: biological and biotechnological tools. *Biotechnol. Advances*, *32*, 73-86.

Parlak, K. U. & Yilmaz, D. D. (2013). Ecophysiological tolerance of *Lemna gibba* L. exposed to cadmium. *Ecotoxicol. Environ. Saf.*, *91*, 79-85.

Pasqualini, S., Piccioni, C., Reale, L., Ederli, L., Della Torre, G. & Ferranti, F. (2003). Ozone-induced cell death in tobacco cultivar Bel W3 plants. The role of programmed cell death in lesion formation. *Plant Physiol.*, *133*, 1122-34.

Quevauviller, P. & Maier, E. A. (1994). *Quality assurance of sampling and sample pre-treatment for soil and sediment analysis*. In European Commission, Environmental Quality of Life. Research Trends in the Field of Environmental Analysis. Results of Scientific Workshop Held within the Measurement and Testing Programme (BCR). Final Report, Directorate General, Science, Research and Development.

Rakhshaee, R., Giahi, M. & Pourahmad, A. (2009). Studying effect of cell wall's carboxyl–carboxylate ratio change of *Lemna minor* to remove heavy metals from aqueous solution. *J. Hazardous Materials*, *163*, 165–173.

Rai, P. K. (2008). Heavy metal pollution in aquatic ecosystems and its phytoremediation using wetland plants: an ecosustainable approach. *Int. J. Phytoremediation*, *10*, 133-160.

Rai, P. K. (2012). An eco-sustainable green approach for heavy metals management: two case studies of developing industrial region. *Environ. Monit. Assess.*, *184*(1), 421-48.

Rascio, N. & Navari-Izzo, F. (2011). Heavy metal hyperaccumulating plants: How and why do they do it? And what makes them so interesting? *Plant Sci.*, *180*, 169-181.

Razinger, J., Dermastia, M., Drinovec, L., Drobne, D., Zrimec, A. & Dolenc Koce, J. (2006). Antioxidative responses of duckweed (*Lemna minor* L.) to short term copper exposure. *Environ. Sci. Pollution Res. International*, *14*, 194-201.

Saygidğer, S. & Doğan, M. (2004). Lead and cadmium accumulation and toxicity in the presence of EDTA in *Lemna minor* L. and *Ceratophyllum demersum* L. *Bull. Environ. Contam. Toxicol.*, *73*, 182-189.

Scandalios, J. G. (2002). The rise of ROS. Trends in Plant Science, *27*, 483-486.

Schneider-Nahum, G. (2000). Metal binding peptides in *Azolla*. PhD Thesis, The Hebrew University of Jerusalem.

Schor-Fumbarov, T., Goldsbrough, P., Adam, Z. & Tel-Or, E. (2005). Characterization and expression of metallothionein gene in the aquatic fern *Azolla filiculoides* under heavy metal stress. *Planta*, *223*, 69-76.

Shang, D. Y., MacDonald, R. W. & Ikonomou, M. G. (1999). Persistence of nonylphenol ethoxylate surfactants and their primary degradation products in sediments from near a municipal outfall in the Strait of Georgia, British Columbia, Canada. *Environ. Sci. Technol.*, *33*, 1366–1372.

Shaw, B. P., Sahu, S. K. & Mishra, R. K. (2004). *Heavy metal induced oxidative damage in terrestrial plants*. In Prasad M.N.V. (Ed.) Heavy metal stress in plants. From biomolecules to ecosystems, Springer, New York, 84-126.

Sood, A., Uniyal, P. L., Prasanna, R. & Ahluwalia, A. S. (2012). Phytoremediation Potential of Aquatic Macrophyte, *Azolla*. *AMBIO*, *41*, 122–137

Tabor, C. F. & Barber II, L. B. (1996). Fate of linear alkylbenzene sulfonate in the Mississippi River. *Environ. Sci. Technol.*, *30*, 161–171.

Tel-Or, E. & Forni, C. (2011). Phytoremediation of hazardous toxic metals and organics by photosynthetic aquatic systems. *Plant Biosystems*, *145*(1), 224-235.

Turner, A. & Xu, J. (2008). Influence of ionic surfactants on the flocculation and sorption of palladium and mercury in the aquatic environment. *Water Res.*, *42*, 318–326.

Verlicchi, P. & Zambello, E. (2014). How efficient are constructed wetlands in removing pharmaceuticals from untreated and treated urban wastewaters? A review. *Science Total Environ.*, 470-471, 1281-1306.

Yang, L. Y., Jiang, L. J., Zhou, Z., Chen, Y. G. & Wang, X. R. (2002). The sedimentation capabilities of hexadecyltrimethylammonium-modified montmorillonites. *Chemosphere, 48*, 461–466.

Ying, G. G. (2006). Fate, behavior and effects of surfactants and their degradation products in the environment. *Environ. International, 32*, 417-431

Wilde, K. L., Stauber, J. L., Markich, S. J., Franklin, N. M. & Brown, P. L. (2006). The effect of pH on the uptake and toxicity of copper and zinc in a tropical freshwater alga (*Chlorella* sp.). *Arch. Environ. Contam. Toxicol.*, *51*, 174-85.

Zhang, Y. & Zhou, J. L. (2005). Removal of estrone and 17b-estradiol from water by adsorption. *Water Res.*, *39*, 3991–4003.

Zhao, J., Davis, L. C. & Verpoorte, R. (2005). Elicitor signal transduction leading to production of plant secondary metabolites. *Biotechnol. Adv.*, *23*, 283–333.

In: Macrophytes
Editor: Raven Capello

ISBN: 978-1-63321-518-4

Chapter 3

VEGETATION STRUCTURE OF LAGOON ENVIRONMENTS AS A RESULT OF SEDIMENT BIOGEOCHEMICAL DYNAMICS AND ORGANIC MATTER LOAD: A REVIEW

M. Lenzi*

Lagoon Ecology and Aquaculture Laboratory (LEALab),
Orbetello Pesca Lagunare srl, Orbetello, Italy

ABSTRACT

Shallow lagoons with low water renewal tend to be eutrophic. Eutrophic conditions may become extreme if there is strong human pressure in the surrounding area. They promote opportunistic macroalgae which can constitute vast dense mats. Subsequent decay of biomass enriches sediment with organic matter and may lead to dystrophic status. During these phases, mineralization processes involving bacterial activities, such as ammonification, nitrification, denitrification and sulphate reduction, occur in sediment. Decay and mineralization of biomass imply release of toxic catabolites and nutrients, the latter supporting new blooms of vegetation. Under extreme conditions, macroalgal blooms repeat with increasing frequency, culminating in a dissipating event, which manifests as an extended dystrophic crisis. Here we report a conceptual model of the sediment energy cycle, from low

* lealab2@hotmail.com.

organic matter content of sediment to accumulation and anoxigenic dissipation, and thence to new accumulation, and so on. At the various stages, different quantities and qualities of nutrients are released from the sediment, creating different environmental conditions and blooms of vegetation that differ in quantity and quality.

LAGOON ENVIRONMENTS

Estuarine and lagoon environments are transition areas between marine and inland water domains. They are characterized by wide changes in the values of chemical and physical water variables, due to the changing mix of freshwater and seawater, relatively shallow depth and intense exchange at the air/water, water/land and water/sediment interfaces.

The great variability characterizes these environments as extreme and structurally stressed (McLusky and Elliott, 2007), thus entailing a reduced number of species, where it is impossible to distinguish natural disturbance from human disturbance, and difficult to establish environmental quality, trophism and past and present effects of human activity, according to the concepts of "environmental homeostasis" and "estuarine quality paradox" (Dauvin, 2007; Elliott and Quintino, 2007).

This changeability involves selection against plant and animal populations. Few species can populate environments with such high nutritional potential, and species that can tolerate such variability proliferate. These features are common to lagoon and estuarine environments around the world. They express a single essential biotic community, the lagoon euryhaline and eurythermal biocenosis (LEE), which is structurally similar everywhere, although the species change from one latitude to another.

Lagoon and estuarine environments change in relation to the tide range. The power of tidal flows establishes the type of marine mouths, whether they tend to close or remain open, and the particle-size distribution in the basin, with clays in deeper central areas when tides are small, and clays in peripheral areas of basins deeper near their mouths, when tidal forces are large (Brambati, 1988).

In their classification of estuaries on the basis of tidal range, McLusky and Elliott (2004) considered non-tidal estuaries as limnetic, and estuaries with tidal range < 2 m as microtidal. In the present review, Mediterranean coastal lagoons that have tidal ranges < 50 cm and oligohaline to hyperhaline salinity are classified as non-tidal to distinguish them from North Adriatic lagoons, which have a connotation of environments much affected by the tides, more aptly described as microtidal. This clarification is not trivial, because the influence of eutrophication on a basin is all the more significant, the less the basin is subject to water replacement. In the case of Venice lagoon, which is a microtidal basin with an average tidal range of about 1 m, Sfirso et al. (1994) revealed that the origins of eutrophication between March and September were mostly marine: high lagoon water replacement by marine water, through tidal flow, allowed entry of nutrient loads from rivers draining into the northern Adriatic. For the Tyrrhenian lagoon of Orbetello, Lenzi et al (2003) estimated that water turnover removed about 67% of nitrogen and 28% of phosphorus from anthropogenic sources. The rest accumulated in the ecosystem, a minor part escaping through the air/water interface and levies (seaweed harvesting, fishing, etc.).

This review only considers environmental conditions linked to poor water exchange and shallow depth, where eutrophication is a natural trend enhanced by anthropogenic inputs. Greater depth can lead to a predominance of phytoplankton blooms, while here I want to dwell on macroalgal development.

Eutrophication and Macroalgal Blooms

In recent decades, estuarine environments all over the world have experienced high human pressure and nutrient loads, leading to eutrophication and proliferation of microalgae and/or macroalgae (Hauxwell and Valiela, 2004). Environmental responses to eutrophication vary according to capacity to buffer the effects (Cloern, 2001). Shallow lagoons with little water exchange are particularly subject to eutrophic stress, which may trigger chain reactions leading to a vegetation shift towards increasingly tolerant species and eventually dystrophic status.

Faster-growing opportunistic macroalgae tolerate indirect effects of eutrophication, such as anoxia, elevated concentrations of ammonia and release of hydrogen sulphide from sediment (Gordon et al., 1981; Peckol and Rivers, 1995a, b; Holmer, 1999). Tolerant to changing environmental conditions, they can develop high-density mats. Because of their high growth rate and uptake of nitrogen (N) and phosphorus (P) (Littler and Littler, 1980; Rosemberg and Ramus, 1984), they remove, store and release many dissolved nutrients (Ryther et al., 1981), also playing the role of nutrient source when they decompose (Sfriso et al., 1987; Hanisak, 1993). Sediment anoxic conditions also produce nutrient release, which promotes new algal blooms

(Lavery and McComb, 1991). Opportunistic macroalgae have different nutrient requirements. Some species are nitrophilous (most *Gracilaria*), others are nitrophilous and tolerate environmental degradation due to organic pollution (many *Ulva* withstand ammonium and hydrogen sulphide) (Haritonidis, 1996). Some Chlorophyceae have a relatively higher requirement for P (*Chaetomorpha linum* and especially certain *Cladophora*) (Gordon et al., 1981; Planas et al., 1996). Lenzi et al. (2003) observed a horizontal distribution of these macroalgae in a eutrophic lagoon, and a succession of species in relation to the gradient of nutrients from anthropogenic sources.

According to Littler and Littler (1980), if different macroalgae adapt to the same environment by different evolutionary pathways, partitioning available resources, then most opportunistic macroalgae presumably have their own strategies for exploiting natural resources. These strategies may produce different environmental conditions, especially in high-density mats. According to Lenzi et al. (2013b), biomass alone is not enough to define the impact and risk of collapse of a mat. Density, which expresses the distribution of biomass in the water column and is related to thallus morphology, provides a criterion for dystrophic risk triggered by decay of the algal mass. Mat density is strongly determined by the type of algal thallus: heavy thalli tend to form compact mat; light thalli tend to distribute more widely in the water column. Paradoxically, mats are self-limiting if they maintain low density, and have more opportunity for development if they reach high densities. The pleustophytic habitus of macroalgae in a lagoon environment is functional to this strategy, because it can produce high-density masses, driven by wind, and create biogeochemical conditions in sediments that determine intense nutrient release.

Macroalgal growth produces competition with seagrass and accelerates the degradation of seagrass meadows (Raffaelli et al., 1998; McGlatery, 2001). Under eutrophic conditions, a vegetation shift at the expense of seagrass meadows is a first event (Duarte 1995; Goodman et al., 1995; Terrados et al., 1999). Seagrasses are considered a valuable component of shallow water environments. For example, they favour aerobic oxidation of the top layer of sediment, transporting oxygen produced by photosynthesis to the rhizosphere and dispersing oxygen into the sediment through root lacunae (Pedersen et al., 1998). Removal of the causes of eutrophication and/or control of its negative consequences improve water and sediment quality and promote seagrass recovery (Vandenbroeck and Ben Charrada, 2001; Plus et al., 2003; Lenzi et al., 2003, 2010). Maintenance of seagrass meadows in coastal environments around the world is now a priority (Barbier et al., 2011; Orth et al., 2006). In a meta-analysis, Thomsen et al. (2012) detected wide variability in the impact of macroalgae on seagrasses, making it difficult to identify a general impact factor. Although direct action of macroalgae on seagrasses and mode of impact may be important, it is clear that the causes of environmental degradation and macroalgal blooms are what must be addressed as a priority. In fact, the removal of environmental stressors may result in a gradual return of meadows and a decrease in macroalgal development, whether the stress factor seem to be invasive macroalgae (Lenzi et al, 2013a) or eutrophic conditions (Lenzi et al., 2010).

Eutrophication is not the only factor that may lead to development of opportunistic algae and regression of seagrass beds. Other factors, such as contaminants and global warming, must also be considered. In years of accentuated water warming, Garcìa et al. (2013) observed increased sulphide intrusion into *Posidonia oceanica* leaves (to which the species is particularly sensitive) due to increased activity of sulphate-reducing bacteria. Global warming could make lagoon conditions even more extreme, although it will probably produce more visible effects in open marine areas, more sensitive to changes in environmental variables, than in lagoon areas with low water turnover, where populations are already structurally connoted to withstand extreme conditions. However, eutrophication remains a very important factor along coastlines and under certain conditions can even lead to algal blooms in offshore marine areas (Liu et al., 2013).

Eutrophication and Organic Matter

Though still debated, Nixon's (1995) definition of eutrophication remains valid, namely, "an increase in the rate of supply of organic carbon to an ecosystem", and not solely the nitrogen and phosphorus levels of a water body. With the main aim of giving more weight to anthropogenic eutrophication, a recent draft of a pan-European framework for assessment of eutrophication published the following definition of eutrophication: "the enrichment of water by nutrients, especially nitrogen and/or phosphorus, and organic matter" (Anderson et al., 2006).

An increase in organic matter (OM) in sediments is recognized as a factor of environmental disturbance that can cause big changes and reduce biodiversity. A classic example of disturbance–response relationship of a zoobenthonic community faced with a gradient of sediment enrichment in OM was proposed by Pearson and Rosenberg (1978). Hargrave et al. (2008) used a nomogram to demonstrate a link between the biogeochemical variables used to measure the relationship between OM load and sulphate reduction on one hand, and changes in marine benthic communities on the other.

Macroalgal proliferation and the formation of thick, high density mats prevent light from reaching under-layer vegetation and can lead to stratification of chemical-physical variables (Krause-Jensen et al., 1996; Lenzi et al., 2013b). Hypoxia and bacterial attack can lead to decay of the under layer of the mat, enriching sediment with OM. Sediment is the substrate of many bacterial processes. Under such conditions, mineralization occurs through anoxigenic processes, first through the respiration of nitrate, then through oxidants such as Fe^{3+} and Mn^{4+}, then through sulphates (Stainer et al., 1988). Sulphate-reducing bacteria produce hydrogen sulphide, which has an acidifying effect and high toxicity on all organisms. H_2S can be removed by buffer mechanisms, such as reaction with iron. The ferric ion (Fe^{3+}) acts directly, oxidizing hydrogen sulphide to elemental sulphur, while the ferrous ion (Fe^{2+}) blocks the acid as ferrous sulphide, an insoluble salt (FeS), which gives anoxic sludge its typical black colour. Anoxic sludge can further react with sulphur to form pyrite (FeS_2), an even more insoluble compound (Berner, 1984; Hijs et al., 2000; Rozan et al., 2002). When available ferric and ferrous ions are exhausted, hydrogen sulphide is released from sediment into the water column (Giordani et al., 1996). The system collapses and fauna die-offs can be evident. This phenomenon is called dystrophic crisis.

Dystrophic crisis is a dissipation process by which an ecosystem, which has reached untenable energy conditions, heals itself. Through dystrophic dissipation, the ecosystem gets rid of excess accumulated energy in sediments (Lenzi et al., 2011, 2012).

Relationship between Organic Matter and Vegetation Structure

Accumulation of OM in sediment, the rate of which depends on the degree of eutrophication, causes release of N and P in variable ratios. N is released as various chemical species, and acid or toxic reducing substances are tolerated to different degrees by the various species of aquatic vegetation.

Breakdown of OM leads to production of ammonium (Marty et al., 1990) and nitrite by ammonification processes, causing a decrease in redox potential and acidification due to CO_2 and H_2S production. Dissolved sulphides (H_2S,

HS^-, S^{2-}), which accumulate in pore water by sulphate respiration, impede the activity of nitrification and denitrification bacteria (Joye and Hollibaugh, 1995), increasing benthic ammonium flow and enhancing primary production in the water column. Reduction of iron and its blockade by H_2S leads to release of orthophosphates, previously bound to ferric oxides-hydroxides, into the water column (Gunnars and Blomqvist, 1997; Golterman, 2001; Rozan et al., 2002).

In essence, many nutrients becomes available to opportunist vegetation that tolerates increasingly extreme conditions. Nutrient releases vary with the quantity and quality of the chemical species (such as N) involved and with the quantity and quality of sedimentary OM.

In the case of shallow water ecosystems with low water turnover, vegetation quality and quantity under eutrophic conditions are controlled by sediment load of OM, which is dissipated during dystrophic events (Lenzi et al., 2011).

Vegetation status substantially reflects the energy accumulated in the top centimetres of sediment, those that conduct biogeochemical exchange with the water column. According to the conceptual model of Figure 1, this causes instability and changes in plant populations, setting up vegetation cycles with extremes that go from low sediment energy levels, after a dystrophic crisis, to high levels, just before the subsequent crisis. This view provides an explanation for the great temporal variations in aquatic vegetation of eutrophic estuaries and lagoons, in line with the broader picture of variations proposed by Duarte (1995) for the transition from oligotrophy to eutrophy in marine environments. The cycle involves a series of increasingly tolerant and opportunistic species as sediment becomes enriched in OM, nutrient release from sediment increases and ammonium prevails over nitrate. A dominance of seagrasses, in the best cases, deteriorates to a dominance of Chlorophyceae under increasingly degraded conditions, via a series of intermediate vegetation stages, each of which has particular conditions of sedimentary OM, nutrient release and N:P atomic ratio (Lenzi et al., 2012). A lagoon ecosystem that has undergone an intense dissipative phenomenon may appear totally devoid of vegetation one or two months after the event; it may be oligotrophic in dissolved nutrients and have relatively high sediment Eh. Sea-lagoon communication, if present, allows recruitment of marine species and opportunistic species. In this first phase, marine species typically develop and grow for months, until they collapse under the effect, for example, of summer warming. However, they disappear more or less gradually as the decay of primary production increases sedimentary OM. The time between one dystrophic event and the next depends on eutrophic status: the faster OM accumulates in sediment, the higher the frequency of dystrophic events. The higher the energy dissipation, the higher the recruitment of species and biodiversity occurring transiently in the ecosystem.

The present conceptual model arises from analysis of a 26-year dataset of macroalgal blooms and dystrophic events in Orbetello lagoon (Lenzi et al., 2011), further supported by apparently anomalous blooms, such as that of *Alsidium corallinum*, a marine Rhodophycea (Lenzi et al., 2012).

Tools for Restoration and Prospects

Solutions for environmental remediation have given priority to engineering approaches, often focusing on increasing water turnover by excavation of canals or marine mouths and pumping of water, and/or to a management approach based on macroalgal harvesting. By contrast, the model outlined above suggests solutions that promote natural mineralization processes and provides a tool for restoration, namely measures that avoid accumulation of OM in sediment and maintain a relatively low value, compatible with acceptable biodiversity levels. Such approach ignores the causes of eutrophication, i.e. nutrients from external sources, intervening downstream of eutrophication. This helps to solve the objective difficulty of stopping nutrient input in heavily populated areas. A direct intervention on sediment having a high organic load, encouraging oxygenic mineralization, can be done in different ways, varying with the type and extent of the areas to be treated: resuspension of sediments (Lenzi, 2010), insufflation of air, sediment removal and so forth.

It is not yet possible to establish a relationship between sediment organic matter load and a precise succession of benthic vegetation. Available datasets are inadequate, the variables are many, and the quality of organic matter must be taken into consideration (Cebrián et al., 1998; Lenzi and Renzi, 2011). However, with sufficient information, it would be possible to develop an index of dystrophic risk, based on the quantity and quality of vegetation (Lenzi et al., 2011) and of OM. The index would be an indicator of the likelihood of an imminent large-scale dystrophic process.

I do not believe that environmental quality indices based on biodiversity (especially phytobenthonic species detected in the ecosystem) can be useful for non-tidal lagoon environments, as suggested in the EU Water Framework Directive 2000/60. If non-tidal lagoons are structurally stressed and if it is impossible to distinguish natural disturbance from human disturbance, then biodiversity cannot be a priority issue to investigate, as the Directive implies, nor can it be a criterion for remediation of degraded environments.

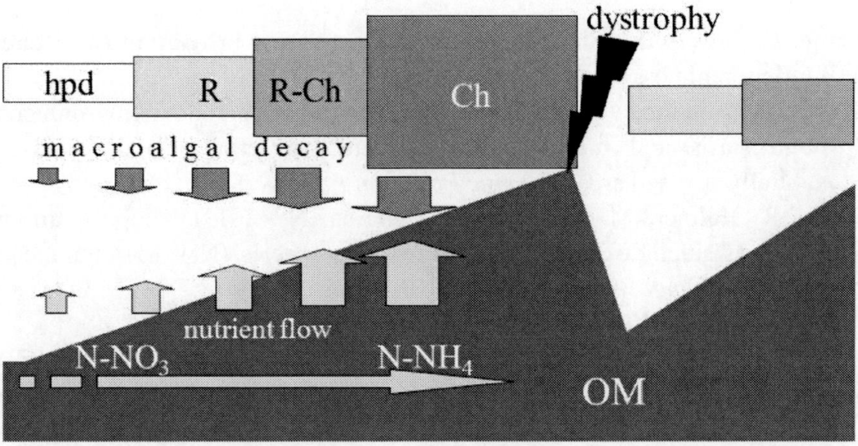

Figure 1. Nutrient release and vegetation dominance alter cycles through accumulation and dissipation of sediment organic matter (dystrophy). Hpd, high biodiversity; R, Rhodophyceae; Ch, Chlorophyceae. From Lenzi et al. (2012).

REFERENCES

Andersen, J.H., Schlűter, L., Ærtebjerg, G. (2006). Coastal eutrophication: recent developments in definitions and implications for monitoring strategies. Journal of plankton research, 28 (7), 621-628.

Barbier, E.B., Hacker, S,D., Kennedy, C.J., Koch, E.W., Stier, A.C. (2011). The Value of Estuarine and Coastal Ecosystem Services. *Ecological Monographs*, 81, 169–193.

Berner, R.A. (1984). Sedimentary pyrite formation: an uptdate. *Geochim. Cosmochim. Acta*, 48, 605-615.

Brambati, A. Lagoons and coastal ponds: two environments in comparison. In: Carrada, G. C., Cicogna, F. & Fresi, E. (Eds.). *Coastal lagoon: Research and Management*. Clem, Massa Lubrense, Napoli, 1988. 215pp, 9-33.

Cebrián, J., Williams, M., McClelland, J., Valiela, I. (1998). The dependence of heterotrophic consumption and C accumulation on autotrophic nutrient concentration in ecosystems. *Ecol. Lett.*, 1, 165–170.

Cloern, J.E. (2001). Our evolving conceptual model of the coastal eutrophication problem. *Marine Ecology Progress Series*, 210, 223-253.

Dauvin, J.C. (2007). Paradox of estuarine quality: benthic indicators and indices, consensus or debate for the future. *Marine Pollution Bulletin*, 55, 271-281.

Duarte, C. M. (1995). Submerged aquatic vegetation in relation to different nutrient regimes. *Ophelia*, 41, 87-112.

Elliott, M., Quintino, V. (2007). The estuarine quality paradox, environmental homeostasis and the difficulty of detecting anthropogenic stress in naturally stressed areas. *Marine Pollution Bulletin*, 54 (6), 640-645.

Garcìa, R, Holmer, M., Duarte, C.M., Marba, N. (2013). Global warming enhances sulphide stress in a key seagrass species (NW Mediterranean). *Global Change Biology*, © 2013 John Wiley & Sons Ltd, doi: 10.1111/gcb.12377

Giordani, G, Batoli, M, Cattadori, M, Viaroli, P. (1996). Sulphide release from anoxic sediments in relation to iron availability and organic matter recalcitrance and its effects on inorganic phosphorus recycling. *Hydrobiologia*, 329, 211-222.

Golterman, H.L. (2001). Phosphate release from anoxic sediments or "what did Mortimer really write?". *Hydrobiologia*, 450, 99-106.

Goodman, J.L., Moore, K.A., Dennison,W.C. (1995). Photosynthetic responses of eelgrass (*Zostera marina* L.) to light and sediment sulfide in a shallow barrier island lagoon. *Aquatic Botany*, 50, 37–47.

Gordon, D.M., Birch, P.B., McComb, A.J. (1981). Effects of inorganic phosphorus and nitrogen on the growth of an estuarine *Cladophora* in culture. *Botanica Marina*, 24, 93-106.

Gunnars, A., Blomqvist, S. (1997). Phosphate exchange across the sediment-water interface when shifting from anoxic to oxic conditions: An experimental comparison of freshwater and brackish-marine systems. *Biogeochemistry,* 37, 203-226.

Hanisak, M.D. (1993). Nitrogen release from decomposing macroalgaes: species and temperature effects. *Journal of Applied Phycology*, 5, 175-181.

Hargrave, B.T., Holmer, M., Newcombe, C.P. (2008). Towards a classification of organic enrichment in marine sediments based on biogeochemical indicators. *Marine Pollution Bulletin*, 56(5), 810–824.

Haritonidis, S. (1996). Effects of Eutrophication: Algal Typology and Community Evaluation. In: Marine Benthic Vegetation. Recent Changes and the Effects of Eutrophication. W. Schramm and P.N. Nienhuis

(Eds.). Ecological studies, vol. 123. Springer-Verlag, Berlin Heidelberg, 414pp.

Hauxwell, J., Valiela, I. (2004). Effects of nutrient loading on shallow seagrass-dominated coastal systems: patterns and processes. In: Nielsen, S.L., Banta, G.T., Pedersen, M.F. (Eds.). *Estuarine Nutrient Cycling: The Influence of Primary Producers*. Kluwer Academic Publishers, Dordrecht, The Netherlands, pp. 59–92.

Hijs, S.K., Azzoni, R., Giordani, G., Jonkers, H.M., Zizzoli, D., Viaroli, P., Gemerden, H. (2000). Sulphide-induced release of phosphate from sediments of coastal lagoons and the possible relation to the disappearance of *Ruppia* sp. *Aquatic Microbial Ecology*, 23, 85-95.

Holmer, M. (1999). The effect of oxygen depletion on anaerobic organic matter degradation in marine sediments. *Estuarine Coastal Shelf Science*, 48, 383–390.

Joye, S.B., Hollibaugh, J.T. (1995). Influence of Sulphide Inhibition of Nitrification on Nitrogen Regeneration in Sediments. *Science*, 270, 623-625.

Krause-Jensen, D., McGlathery, K., Rysgaard, S., Christensen, P.B. (1996). Production within dense mats of the filamentous macroalga *Chaetomorpha linum* in relation to light and nutrient availability. *Marine Ecoloy Progress Series*, 134, 207–216.

Lavery, P.S., McComb, A.J. (1991). The nutritional ecophysiology of *Chaetomorpha linum* and *Ulva rigida* in Peel Inlet, Western Australia. *Botanica Marina*, 34, 251–260.

Lenzi, M. (2010). Resuspension of Sediment as Method for Managing Shallow Eutrophic Lagoon. *Journal of Ecology and the Natural Environment*, 2 (11), 220-234. ISSN 2006-9847 © 2010. http://academicjournals.org/jene

Lenzi, M., Birardi, F., Calzolai, R., Finoia, M.G., Marcone, F., Nocciolini, S., Roffilli, R., Sgroi, S., Solari, D. (2010). Hypertrophic lagoon management by sediment disturbance. *Marine Pollution Bulletin* (Issue 4–6, Estuarine Ecosystems: Structure, Function and Management – ECSA-42 Symposium in Russia), 61, 189–197.

Lenzi, M., Birardi, F., Finoia, M.G. (2013a). Can the marine ecosystem of a *Posidonia oceanica* barrier reef react and defend itself against the spread of *Caulerpa racemosa* var. *cylindracea* ? *International Journal of Marine Science*, 3 (20), 158-165.

Lenzi, M., Gennaro, P., Renzi, M., Persia, E., Porrello, S. (2012). Spread of *Alsidium corallinum* C. Ag. in a Tyrrhenian eutrophic lagoon dominated by opportunistic macroalgae. *Marine Pollution Bulettin*, 64(12), 2699-2707. doi: 10.1016/j.marpolbul.2012.10.004.

Lenzi, M., Gennaro, P., Mercatali, I., Persia, E., Solari, J., Porrello, S. (2013b). Physico-chemical and nutrient variable stratifications in the water 4 column and in macroalgal thalli as a result of high biomass mats 5 in a non-tidal shallow-water lagoon. *Marine Pollution Bulletin*, 75, 98-104

Lenzi, M., Renzi, M. (2011). Effects of artificial disturbance on quantity and biochemical composition of organic matter in sediments of a coastal lagoon. *Knowledge and Management of Aquatic Ecosystems*, 402, 08, DOI: 10.1051/kmae/2011058

Lenzi, M., Renzi, M., Nesti, U., Gennaro, P., Persia, E., Porrello, S. (2011). Vegetation cyclic shift in eutrophic lagoon. Assessment of dystrophic risk indices based on standing crop evaluation. *Estuarine, Coastal and Shelf Science*, doi:10.1016/j.ecss.2011.10.006

Lenzi, M., Palmieri, R., Porrello, S. (2003). Restoration of the eutrophic Orbetello lagoon (Tyrrhenian Sea, Italy): water quality management. *Marine Pollution Bulletin*, 46, 1540–1548.

Littler, M.M., Littler, D.S. (1980). The evolution of thallus form and survival strategies in benthic marine macroalgae: field and laboratory tests of a functional form model. *American Naturalist*, 116 (1), 25-44.

Liu, D., Keesing, J.K., He, P., Wang, Z., Shi, Y., Wang, Y. (2013). The world's largest macroalgal bloom in the Yellow Sea, China: Formation and implications. *Estuarine, Coastal and Shelf Science*, 129, 2-10.

Marty, D., Esnault, G., Caumette, P., Ranaivoson-Rambeloarisoa, E., Bertrand, J.C. (1990). Denitrification, sulfato-reduction et methanogenese dans les sediments superficiels d'un étang saumatre méditerranéen. *Oceanologica Acta*, 13, 199-210.

McGlathery, K. (2001) Macroalgal blooms contribute to the decline in seagrasses in nutrient-enriched coastal waters. *Journal of Phycology*, 37, 453–456.

McLusky, D.S., Elliott, M. The Estuarine Ecosystem. Ecology, Threats, and Management. Oxford University press, 2004, 214pp.

Nixon, S.W. (1995). Coastal marine eutrophication: a definition, social causes and future concerns. *Ophelia*, 41, 199-219.

Orth, R., Carruthers, T.J.B., Dennison, W.C., Duarte, C.M., Fourqurean, J.W. (2006). A global crisis for seagrass ecosystems. *BioScience*, 56, 987–996.

Pearson, T.H., Rosenberg, R. (1978). Macrobenthic succession in relation to organic enrichment and pollution of the marine environment. *Oceanography and Marine Biology: An Annual Review*, 16, 229–311.

Peckol, P., Rivers, J.S. (1995a). Physiological responses of the opportunistic macroalgae *Cladophora vagabunda* (L.) van den Hoek and *Gracilariopsis tikvahiae* (McLachlan) to environmental disturbances associated with eutrophication. *Journal of Experimental Marine Biology and Ecology*, 190, 1-16.

Peckol, P., Rivers, J.S. (1995b). Competitive interactions between the opportunistic macroalgae *Cladophora vagabunda* (Chlorophyta) and *Gracilariopsis tikvahiae* (Rodophyta) under eutrophic conditions. *Journal of Phycology*, 31, 229-232.

Pedersen, O., Borum, J., Duarte, C.M., Fortes, M.D. (1998). Oxygen dynamics in the rhizosphere of *Cymodocea rotundata*. *Marine Ecology Progress Series*, 169, 283–288.

Planas, D., Maberly, S.C., Parker, J.E. (1996). Phosphorus and nitrogen relationship of *Cladophora glomerata* in two lake basins of different trophic status. *Freshwater Biology*, 35, 609–622.

Plus, M., Dessous-Paoli, J.-M., Dagoult, F. (2003). Seagrass (*Zostera marina* L.) bed recolonisation after anoxia-induced full mortality. *Aquatic Botany*, 77, 121–134.

Raffaelli, D.G., Raven, J.A., Poole, L.J. (1998). Ecological impact of green macroalgal blooms. *Oceanography and Marine Biology: an Annual Review*, 36, 97–125.

Rosemberg, G., Ramus, J. (1984). Uptake of inorganic nitrogen and macroalgae surface area: volume ratio. *Aquatic Botany*, 19, 65-72.

Rozan, T.F., Taillefert, M., Trouwborst, R.E., Glazer, B.T., Ma, S., Herszage, J., Valdes, L.M., Price, K.S., Luther, G.W. (2002). Iron-sulphur-phosphorus cycling in the sediments of a shallow coastal bay: Implications for sediment nutrients release and benthic macroalgal blooms. *Limnology and* Oceanography, 47, 1346-1354.

Ryther, J.H., Corwin, N., DeBusk, T., Williams, L.D. (1981). Nitrogen uptake and storage by the red alga *Gracilariopsis tikvahiae* (McLachlan 1979). *Aquaculture*, 26, 107-115.

Sfriso, A., Marcomini, A., Pavoni, B. (1987). Relationships Between Macroalgal Biomass and Nutrient Concentrations in a Hypertrophic Area of the Venice Lagoon. *Marine Environmental Resources*, 22, 297-312.

Sfriso, A., Marcomini, A., Pavoni, B. (1994). Annual nutrient exchanges between the central lagoon of Venice and the northern Adriatic sea. *Science of The Total Environment*, 156 (77), 92. DOI:10.1016/0048-9697(94)90421-9

Stanier, R.Y., Ingraham, J., Wheelis, M.L., Painter, P.R. The world of micro-organisms. Zanichelli, 1988. 676pp. ISBN: 9788808064561

Terrados, J., Duarte, C.M., Kamp-Nielsen, L., Agawin, N.S.R., Gacia, E., Lacap, D., Fortes, M.D., Borum, J., Lubanski, M., Greve, T. (1999). Are seagrass growth and survival constrained by the reducing conditions of the sediment? *Aquatic Botany*, 65, 175–197.

Thomsen, M.D., Wernberg, T., Engelen, A.H., Tuya, F., Vanderklift, M.A., Holmer, M., McGlathery, K.J., Arenas, F., Kotta, J., Silliman, B.R. (2012). A Meta-Analysis of Seaweed Impacts on Seagrasses: Generalities and Knowledge Gaps. *PLoS ONE* www.plosone.org, 7 (Issue 1) e28595.

Vandenbroeck, J., Ben Charrada, R. (2001). Restoration and development project of South Lake of Tunis and its shores. *Terra et Aqua*, 85, 11–20.

In: Macrophytes
Editor: Raven Capello
ISBN: 978-1-63321-518-4

Chapter 4

MACROPHYTES IN NEOTROPICAL AQUACULTURE FARMS: POSITIVE AND NEGATIVE EFFECTS

Lúcia Helena Sipaúba-Tavares**, Samuel Generoso Dias, Fernanda Travaini-Lima, Bruno Scardoeli-Truzzi and Cecília Anatriello Boareto
São Paulo State University, Aquaculture Center (CAUNESP), Limnology and Plankton Production Laboratory, Road Prof. Paulo Donato Castellane s/n, Jaboticabal, SP, Brazil

ABSTRACT

Aquatic plants play a significant role in maintaining water quality, due to their ability of absorbing nutrients. The aim of this chapter is to show the value of macrophyte use as a tool for biological water treatment of aquaculture farms and also as a culture medium for microalgae. In southeastern Brazil, aquaculture farming is mainly carried out in flow-through systems, in which large earthen fishponds are commonly arranged in line, each pond receiving the effluent from the previous one, without any prior treatment. The continuous water flow through the fishponds strongly depends on seasonal rains. The current system for production of aquatic organisms could benefit greatly from the development of low cost strategies to mitigate its impacts in the

* Corresponding author: e-mail:sipauba@caunesp.unesp.br.

environment. Maintenance of good water quality is of primary importance to maintain an ideal culture ecosystem, with adequate feed for optimal fish yield potential, and preservation of a eutrophic situation. Fish production may cause serious environmental issues, especially when there is little concern over excessive plant growth, which can influence the functioning and structure of ecological processes in the fishponds. The use of macrophytes as bio-filter and / or wetland or as microalgae culture medium in aquaculture farms can serve as a tool for maintenance of good environmental conditions and improve the cultivation of microalgae. These microalgae can in turn be used as a nutritional source for the fish larvae, pigment in aquaculture and also as a supplementary additive in the food industry. This alternative medium should allow efficient production of microalgae, at a low cost for the laboratory. The understanding of local characteristics and development of proper feeding techniques, aimed at the integration and optimization of fish production, are important factors in obtaining and maintaining water quality appropriate for microalgae culture.

INTRODUCTION

Owing to increasingly intensive and semi-intensive aquaculture, high amounts of fertilizers are used in fish culture ponds nowadays. The fish do not utilize large quantities of these elements and accumulations occur. Consequently, most fishponds become hypernutrified and undergo environmental degradation problems such as algal blooms, ammonia toxicity, fish diseases and macrophyte infestations, due to high levels of organic matter and nitrogen. The above serious negative effects cause severe economic losses to aquaculture, fisheries and mostly, to the environment (Ferdoushi et al., 2008).

The excess of nutrients in the fishponds can be used to grow aquatic macrophytes, which will contribute in the preservation of desirable water quality for fish farming. Aquatic plants are highly capable of removing nutrients from the water column or the sediment during their growth process. In fact, they partially reduce the resulting load and improve water quality. Appropriate management knowledge and control of the environmental and biological aspects of aquaculture are important to obtain and maintain proper water and sediment quality, which guarantees high biomass production. When water is drained directly from one pond to another, it may affect the characteristics of the subsequent fishponds. In such cases, macrophytes may be a good alternative treatment to remove nutrients from fishponds in this continuous water flow system (Van Dam et al., 2002). Aquatic plants can act as measurable indicators of the ecological conditions of surface waters. Floating macrophytes take up inorganic nutrients mainly by the roots, though uptake through the leaves may also be significant. Improvement of water quality in eutrophicated fishponds can be achieved with the use of floating macrophytes, such as *Eichhornia crassipes* (Sipaúba-Tavares et al., 2003).

Certain inconveniences in the use of water plants may occur. For instance, excessive growth and natural death with a consequent rise in nutrient rates by decomposition. Although macrophytes in fish ponds may occupy only 10% of the water surface, when this occupation exceeds 50% or even becomes total, the effect turns out to be adverse. The disadvantage may happen not only because of the lack of light penetration into the water, but also because of the drastic reduction of dissolved oxygen and excessive increase in organic matter in the sediment (Sipaúba-Tavares et al., 2003).

The use of biological treatments, such as constructed wetlands, may be a good tool to remove pollutants from contaminated aquaculture effluents. These systems are based on natural wetlands with aquatic plants in the substrata, such as sand, pebbles or other inert material, which enhance the growth of various communities of microorganisms (Kadlec & Wallace 2009).

Appropriate interactions between plants, soil and microorganisms will reduce the amount of organic and inorganic compounds in the environment and remove unwanted microorganisms, such as *Escherichia coli* and other thermotolerant bacteria, that are hazardous to human health (Ulrich et al., 2005). These systems may influence nutrient cycling and can improve water quality by reducing water velocity, which allows for the retention of suspended particles, and also via a direct uptake of nutrients, thereby providing support for microbial communities, such as nitrifying and denitrifying bacteria (Chen et al., 2010). In addition, macrophytes may impact several other physical and chemical properties of the water column, such as dissolved oxygen, inorganic carbon, pH and alkalinity, as a result of their metabolism (Caraco & Cole, 2002). Constructed wetlands, which are easily maintained low-cost systems, may be a better alternative for the treatment of effluents in developing countries with serious concerns regarding water eutrophication (Sipaúba-Tavares et al., 2002).

Studies on the performance of water plant based treatments in aquaculture effluents have showed that plants are highly efficacious in the removal of nitrogen, phosphorus, BOD_5, and also of suspended solids and thermotolerant coliforms, when compared to systems without any vegetation (Sipaúba-Tavares et al., 2002; Henry-Silva & Camargo, 2006; Travaini-Lima &

Sipaúba-Tavares, 2012; Millan et al., 2014). Plants stimulate nutrient removal in wetlands and, although not fully understood, the use of a variety of different species may be a positive factor in the efficiency of the system. In fact, each species may have different responses to seasonal changes, with great overall improvement throughout the year (Maltais-Landry et al., 2009).

Temperate wetlands are relatively well studied, in comparison to tropical and neotropical wetlands, which have received less attention from the scientific and management communities thus far. Recommendations developed in temperate wetlands may not be directly applicable in humid tropical and neotropical areas because of extreme hydrological inputs (i.e. high annual precipitation), and dissimilar processes and interactions among ecosystem components (Kaplan et al., 2011).

Aquatic environments with low nutrient content usually have small plants as predominant vegetation. Macrophytes are effective in treating organic matter, nitrogen, and phosphorus and also in reducing concentrations of heavy metals, organic chemical and pathogens in shallow reservoirs (Zhang et al., 2011). Since there is in Brazil an insufficient amount of data about the use of macrophytes in shallow water aquaculture reservoirs, further evaluations are necessary to determine the effectiveness of the macrophytes as tools in the maintenance of good conditions in the water column of aquaculture farms (Dias & Sipaúba-Tavares, 2012).

In regards to water plants, efforts are made to obtain new combinations that prove to be efficient as a medium for the microalgae, promoting high density counts and nutritional values. Microalgae are widely used in aquaculture as natural food or as an additive, to improve the growth and survival of small fish. However, in order to be used in aquaculture, the microalgae strains have to meet various criteria, such as ease of culturing, lack of toxicity, high nutritional value and low cost of production. Macrophyte medium is a low-cost fertilizer, and it is a good tool in the microalgae culture, mainly Chlorophyceae species (*Ankistrodesmus gracilis*) (Sipaúba-Tavares et al., 2009).

The production systems of aquatic organisms can benefit greatly of the development of low cost strategies to mitigate the impacts caused by aquaculture farming. Such developments can also result in significant water savings for the fish farmers. Maintenance of good water quality is of primary importance, so that excellent culture conditions are attained and eutrophic status is monitored, to optimize fish yield.

MACROPHYTES AND NEOTROPICAL AQUACULTURE FARMS

Fishponds are environments completely managed and handled by man, with the regular addition of fertilizers in order to increase the productivity and profitability of the system. The exploitation of such environments in Brazil has been carried out in the absence of sufficient limnological knowledge of fishponds. Since there has been an increase in the number of aquaculture farms in Brazil in recent years, many aquaculture farms have been regarded as research models by governmental departments. Most aquaculture farms receive water directly from rivers or reservoirs and, when poor water management is employed, discharge effluents back in these environments without any prior treatment. Most aquaculture farms are divided into compartments, such as laboratories for larvae culture, nutrition fishponds, plankton production ponds (live food), fish fattening ponds, fresh water shrimp cultures, frog farms and others. Due to inappropriate fertilization and fish feeding control, the above-mentioned compartments may strongly deteriorate the water quality. Aquaculture farms are influenced by complex physical and chemical factors and by biological interactions, which depend directly on water quality.

Limnological studies are important to understand aquatic environments, providing the basis for the adequate use of water. They provide data that can contribute to the development of technologies for water management and guarantee water supply in suitable conditions for aquaculture.

Oscillation of physical and chemical parameters in fishponds depend essentially on energy input, with special emphasis on the frequency and nature of the nutrient discharge added to the system. In general, aquaculture farms receive adequate water availability and constant renewal supply but the water is discharged into the river without any treatment (Sipaúba-Tavares et al., 2010). There are two major factors that influence water quality in aquaculture farms: food supplied to fish and fertilization. In fact, 75% of the food given to fish is dejected into the environment as urine (nitrogen) and feces (phosphorus). Fertilization is one of the most important problems concerning fishponds, or rather, the determination of the best fertilizer doses to be added to the pond. The decisions regarding appropriate stock density, water flow, quality and quantity of diet are of great environmental importance and essential components of management. These factors can be a source of environment degradation when posited without any quality control (Eler et al., 2006).

Continuous water renewal is another method to diminish deposition of material on the pond bottom and to lessen the effects of eutrophication. However, continuous flow may bring more serious problems in subsequent ponds and to the local of water discharge. In fact, water excessively rich in nutrients, organic matter and suspended solids provokes eutrophication and sedimentation in natural water bodies due to accumulation of these components (Boyd & Queiroz, 2001). Proper management techniques, notably regarding water discharge in acceptable conditions, may improve water quality, lowering stress on cultivated organisms and increasing the survival and growth of fish (Macedo & Sipaúba-Tavares, 2005).

In southeastern Brazil, fish culture is mainly carried out in flow-through systems, in which large earthen fishponds are commonly lined up in sequence, each pond receiving the effluents from the previous one without any previous treatment. Since each pond receives a daily feed, the accumulating effect through the line of ponds promotes an increase in the water nutrient load. The continuous water flow through the fishponds strongly depends on seasonal rains. Appropriate management, knowledge of local inputs and development of feeding techniques that are based on the concept of integration as means for optimization of fish production are important to obtain and maintain proper water quality.

In the neotropical aquaculture farm (Figure 1) of São Paulo State University (UNESP, Brazil, 21°15'S and 48°18'W), the first fishpond is a water-supply reservoir to the aquaculture farm, with continuous water flow surface area of 3,472 m^2 and water depth ranging from 1 to 1.3 m, located close to an agricultural area (Figure 1-C). The shallow water-supply reservoir is currently almost covered by macrophytes, mainly *Eichhornia azurea* (Sw.) Kunth and *Salvinia auriculata* Aublet. While the former is a perennial floating aquatic plant typically rooted in the mud, the latter is a free-floating aquatic macrophyte, annual or perennial, with small-size leaves. This shallow reservoir was in the past used as a fishpond, but since 1995 is used only as a water reservoir for the aquaculture farm. It is greatly affected by continuous water flow, wind, rainfall, transport nutrients and allochthonous materials that alter the physical and chemical characteristics of the water column.

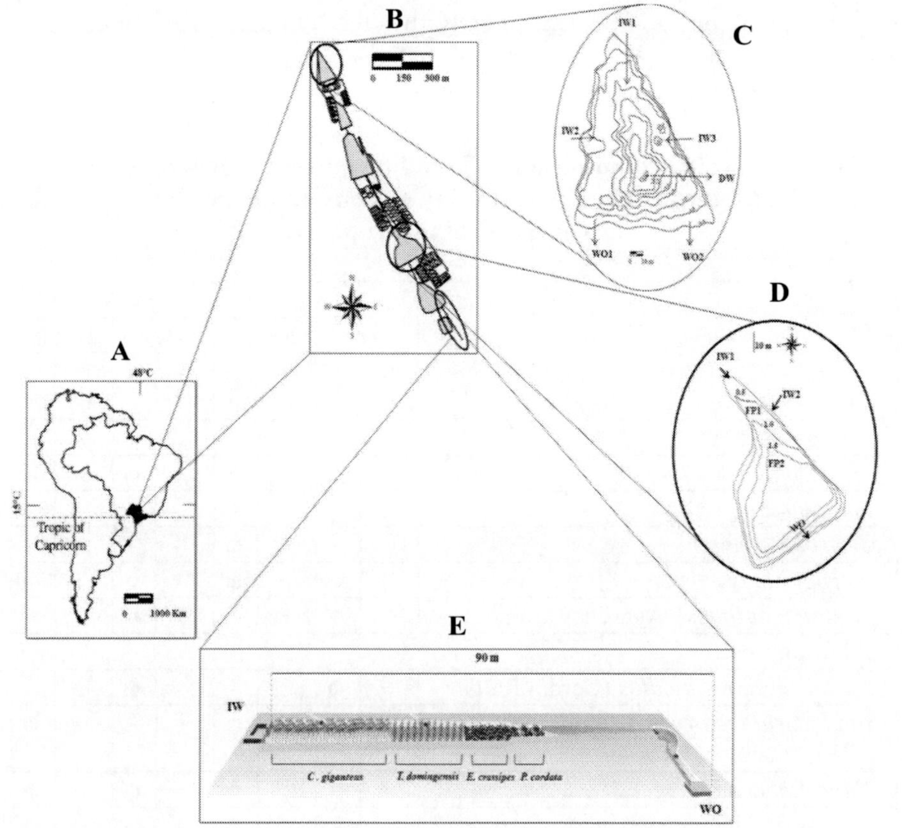

Figure 1. Aquaculture farm: Inset A: shaded area indicates São Paulo state, in southeastern Brazil. Inset B: distribution of ponds at UNESP farm. Inset C: shallow water supply reservoir. Inset D: fishpond infested with macrophytes. Inset E: constructed wetland. IW = inlet water; WO = water outlet; DW = deep water; FP1 and FP2 = sites inside of fishpond.

The macrophytes in the shallow water supply reservoir work as a tool for maintenance of good water conditions by performing a bio-filtration of organic and inorganic pollutants. Rainfall affects the inflow of nutrients to this water supply reservoir, increasing the environmental trophic load and impairing the quality of water supply for the aquaculture system. Very distinct conditions between dry and rainy seasons were observed, in relation to the planktonic community, which showed a low number of species in the rainy season (Tables 1 and 2).

The plankton is affected directly by the climatic and limnological variables, which are factors that can induce significant alterations in the environment and its communities (Perbiche-Neves et al., 2011).

Table 1. Specific composition (SC) and frequency of ocurrence (F) of phytoplanktonic taxa in dry and rainy seasons, in the presence (M) and absence (W) of macrophytes

Taxa	M				W			
	Dry		Rainy		Dry		Rainy	
	SC	F	SC	F	SC	F	SC	F
Cyanobacteria								
Dictyosphaerium sp.	+	○	-	-	+	ф	-	-
Limnothrix sp.	-	-	+	●	-	-	-	-
Merismopedia sp.	-	-	+	ф	-	-	-	-
Microcystis sp.	+	○	+	ф	+	○	+	ф
Oscillatoria limosa Agardhex Gomont	+	○	+	○	+	●	+	●
Chlorophyceae								
Ankistrodesmus falcatus (Corda) Ralfs	+	ф	-	-	+	ф	-	-
Binuclearia tectorum (Kützing) Berger ex Wichmann	+	○	+	ф	-	-	+	ф
Chlorella sp.	+	○	+	○	+	ф	+	ф
Pediastrum duplex Meyen	-	-	+	ф	+	ф	+	ф
Tetraëdron trigonum (Nägeli) Hansgirg	-	-	+	ф	+	ф	+	ф
Zygnematophyceae								
*Bambusina*sp.	+	ф	+	●	+	ф	+	●
Closterium ehrenbergii MeneghiniexRalfs	+	●	+	○	+	ф	+	ф
Closterium setaceum Ehrenberg ex Ralfs	+	○	+	ф	+	ф	-	-
Cosmarium contractum O.Kirchner	+	○	+	○	+	ф	-	-
Cosmarium pyramidatum BrébissonexRalfs	+	○	+	ф	+	ф	-	-
Cosmarium subcostatum Nordstedt	-	-	+	ф	-	-	+	ф
Desmidium sp.	+	ф	+	○	+	○	+	ф
Euastruma bruptum Nordstedt	+	○	+	●	+	○	+	ф
Euastrum subintegrum Nordstedt	+	ф	+	●	+	ф	+	ф

Taxa	M				W			
	Dry		Rainy		Dry		Rainy	
	SC	F	SC	F	SC	F	SC	F
Gonatozygon sp.	+	○	+	ϕ	+	ϕ	-	-
Micrasteria furcata C. Agardhex Ralfs	+	○	+	●	+	○	+	ϕ
Micrasterialaticeps Nordstedt	+	○	+	●	+	○	+	○
Mougeotia sp.	+	○	+	○	+	○	+	ϕ
Netrium sp.	+	○	+	ϕ	+	ϕ	+	○
Pleurotaenium trabecula (Ehrenberg) Nägeli	+	○	+	○	+	●	+	○
Spirogyra majuscula Kützing	+	○	+	○	+	○	+	ϕ
Spondylosium sp.	+	○	+	●	+	●	+	○
Staurastrum rotula Nordstedt	+	●	+	ϕ	+	○	+	○
Staurastrum setigerum Prescott, Bicudo e Vinyard	+	ϕ	+	●	+	○	+	○
Staurodesmus brevispina Croasdale	+	○	+	ϕ	+	ϕ	-	-
Staurodesmus validus Thomasson	+	ϕ	+	○	+	○	+	ϕ
*Zygnema*sp.	+	○	+	ϕ	+	●	+	○
Bacillariophyceae								
Asterionella formosa Hassall	+	○	+	ϕ	+	○	+	ϕ
Eunotia anamargaritae D. Metzeltin Lange-Bertalot	+	●	+	●	+	○	+	ϕ
Eunotia sp.	+	●	+	●	+	●	+	○
Melosira sp.	+	●	+	●	+	●	+	○
Navicula sp.	+	○	+	○	+	●	+	○
Nitzschia amphibia Grunow	+	●	+	●	+	○	+	○
Pinnularia sp.	+	●	+	●	+	●	+	●
Surirella sp.	+	○	+	●	+	●	+	ϕ
Oedogoniaceae								
Oedogonium sp.	+	●	+	●	+	○	+	○
Microsporaceae								
Microspora sp.	+	○	+	●	+	●	+	○

+ = presence; - = absence; ● = constant; ○ = common and ϕ = rare.

In the shallow reservoir of the studied aquaculture farm, the presence of the macrophytes *E. azurea* and *S. auriculata* created favorable conditions to phytoplankton and zooplankton species. Since macrophytes slowed down the water flow, more constant plankton species were observed (Tables 1 and 2). In the water supply reservoir, the phytoplankton community consisted of 42 taxa, with the Zygnematophyceae group being the most representative, followed by Bacillariophyceae, Chlorophyceae, Cyanobacteria, Oedogonophyceae and Microsporaceae (Table 1). The density of Bacillariophyceae was higher in macrophyte-containing areas in both seasons, whereas in sites without macrophytes (primarily water outlet sites), a high density of Cyanobacteria (mainly *Oscillatoria limosa*) was observed. *Chlorella* sp. was the most frequent species (Table 1).

The macrophyte stands support epiphytic algae, as well as a variety of associated mobile animals, including zooplankton, macrofauna and fish (Albertoni et al., 2007; Christie et al., 2009). Macrophytes provide habitat and refuge for zooplanktonic filter feeders to maintain top-down control on the phytoplankton (Sandilands & Hann, 1997). Macrophytes play a central role in the control of phytoplankton and sustain a high faunal diversity. The algae species were correlated with the *E. azurea* and *S. auriculata* biomass.

The zooplankton community consisted of 22 taxa, more specifically 18 Rotifera, 2 Cladocera and 2 Copepoda. The most dominant Rotifera genus were *Lecane*, *Proales*, *Proalinopsis* and *Colurella* (Table 2).

The Rotifera in this study showed higher abundance and diversity (Table 2). Taxa in this group are predominant in environments with human activities, due to their rapid reproduction and adaptation to ecological changes (Dahms et al., 2011). The dominance of the Rotifera over Copepoda and Cladocera genera reflects an exploitative competition between the two zooplanktonic groups. This is also connected with the presence of different forms of algae, which constitute a food base for zooplankton. The structural complexity of aquatic vegetation is likely to provide a wide variety of potential havens for Crustacea predators.

Submerged macrophytes are considered most suited as a daytime refuge, although some studies indicate the occurrence of a protective potential offered by stands of floating leaves and emergent vegetation, functioning as a shelter to zooplankton community (Onsem et al., 2010).

The obtained results show that the presence of macrophytes in a shallow reservoir can be favorable to the maintenance of a diverse plankton community. This study confirmed the important role of macrophytes not only in stabilizing the clear-water state and maintaining high diversity of the plankton community, but also as a viable option to help maintain a desirable quality for the water supply used in aquaculture farm.

Table 2. Specific composition (SC) and frequency of ocurrence (F) of zooplanktonic taxa in dry and rainy seasons, in the presence (M) and absence (W) of macrophytes

Taxa	M				W			
	Dry		Rainy		Dry		Rainy	
	SC	F	SC	F	SC	F	SC	F
Copepoda								
Argyrodiaptomus furcatus Sars, 1901	+	●	+	●	+	●	+	○
nauplii	+	●	+	●	+	●	+	●
Thermocyclops decipiens Kiefer, 1929	+	●	+	●	+	●	+	○
nauplii	+	○	+	●	+	●	+	●
Cladocera								
Alona monacantha Sars, 1901	+	○	+	○	+	●	+	●
Diaphanosoma birgei Korinek, 1981	+	●	+	●	+	●	+	Φ
Rotifera								
Asplanchna sp.	+	ϕ	+	ϕ	+	ϕ	+	Φ
Brachionus quadridentatus Hermann, 1783	+	●	+	ϕ	+	ϕ	+	Φ
Cephalodella sp.	-	-	+	ϕ	+	ϕ	-	-
Colurella dicentra Gosse, 1887	+	ϕ	+	ϕ	+	ϕ	-	-
Colurella sp.	-	-	-	-	+	ϕ	-	-
Lecane bulla Gosse, 1851	+	●	+	ϕ	+	ϕ	+	Φ
Lecane curvicornis Murray, 1913	+	ϕ	-	-	+	ϕ	+	Φ
Lecane lunares Ehrenberg, 1832	+	ϕ	-	-	+	ϕ	-	-
Lecane quadridentata Ehrenberg, 1830	+	ϕ	+	ϕ	+	ϕ	+	Φ
Lecane sp.	+	ϕ	+	ϕ	+	ϕ	+	Φ
Lecane stichaea Harring, 1913	+	ϕ	+	ϕ	-	-	+	Φ
Lepadella ovalis Muller, 1896	+	ϕ	+	ϕ	+	ϕ	-	-
Macrochaetus altamirai Arévalo, 1918	+	ϕ	-	-	+	ϕ	-	-
Monommata longiseta Müller, 1786	+	●	+	ϕ	+	ϕ	+	Φ
Proales doliaris Rousselet, 1895	+	●	+	ϕ	+	○	+	Φ
Proales sp.	+	ϕ	+	ϕ	+	ϕ	-	-
Proalinopsis caudatus Collins, 1872	+	ϕ	+	ϕ	+	ϕ	-	-
Trichocerca longiseta Schrank, 1802	+	●	+	ϕ	+	ϕ	+	Φ

+ = presence; - = absence; ● = constant; ○ = common and ϕ = rare.

Normally, the construction of wetlands prioritizes a reduction in the impact of wastewater generation by aquaculture. In these systems, it is possible to gather quantitative data about of the inputs, outputs and internal cycling of materials during the different seasons (dry and rainy). Such data are important to better characterize how the ecosystem works as a source, a recipient and as a modifier of the chemical compounds. The use of macrophytes in the treatment of wastewater has been the object of great interest in Brazil (Sipaúba-Tavares et al., 2002; Souza et al., 2004; Santiago et al., 2005; Henry-Silva et al., 2008; Travaini-Lima & Sipaúba-Tavares, 2012; Millan et al., 2014).

This study used a wetland environment (90 m^2), that was created in the effluent of the aquaculture farm at UNESP. These studies used three macrophyte species that are predominant in the region: *Cyperus giganteus* (20 m^2), *Typha domingensis* (13 m^2), *Eichhornia crassipes* (10 m^2) and *Pontederia cordata* (6 m^2) (Figure 1-E).

The highest removal efficiency rates of the macrophytes were observed during the dry season, when the air temperature, rainfall and hydraulic loading rates were lower, and the retention time and vegetation biomass were higher. The removal rates for nitrate (56 - 82%), nitrite (66 - 87%) and ammonia (76 - 81%) during the dry season were in general two times higher than these same chemicals during the rainy season (63 - 64%; 29 - 46% and 31 - 65%, respectively). Additionally, the removal rates of SRP (71 - 81%), TP (79 - 86%) and BOD_5 (66 - 92%) were also higher during the dry season, with the mean removal percentage around 84%. In the rainy season, the mean removal percentage was of 55% (Figure 2).

During the rainy season, the water outlet presented a greater concentration of NH_4, NO_2 and NO_3 when compared to the water inlets, indicating that the balance between the nitrification/denitrification was exceeded. The removal of phosphorus from the water systems occurred through biotic processes, such as assimilation by microbial, phytoplankton and plant biomass, and also through abiotic processes such as oxidation, sedimentation, adsorption and precipitation by ions.

The highest removal rates of nitrogen and phosphorus compounds that were observed during the first 20 days after the insertion of plants in the constructed wetland system can be explained by the quick growth of the macrophyte biomass (Figure 2, Table 3).

During the dry season, the abundant presence of vegetal biomass probably played a major role in yielding the highest removal rates of nitrogen and phosphorus from the wetland systems. Corroborating this hypothesis, it was observed that the concentration of nitrogen and phosphorus in the planted area was higher than average.

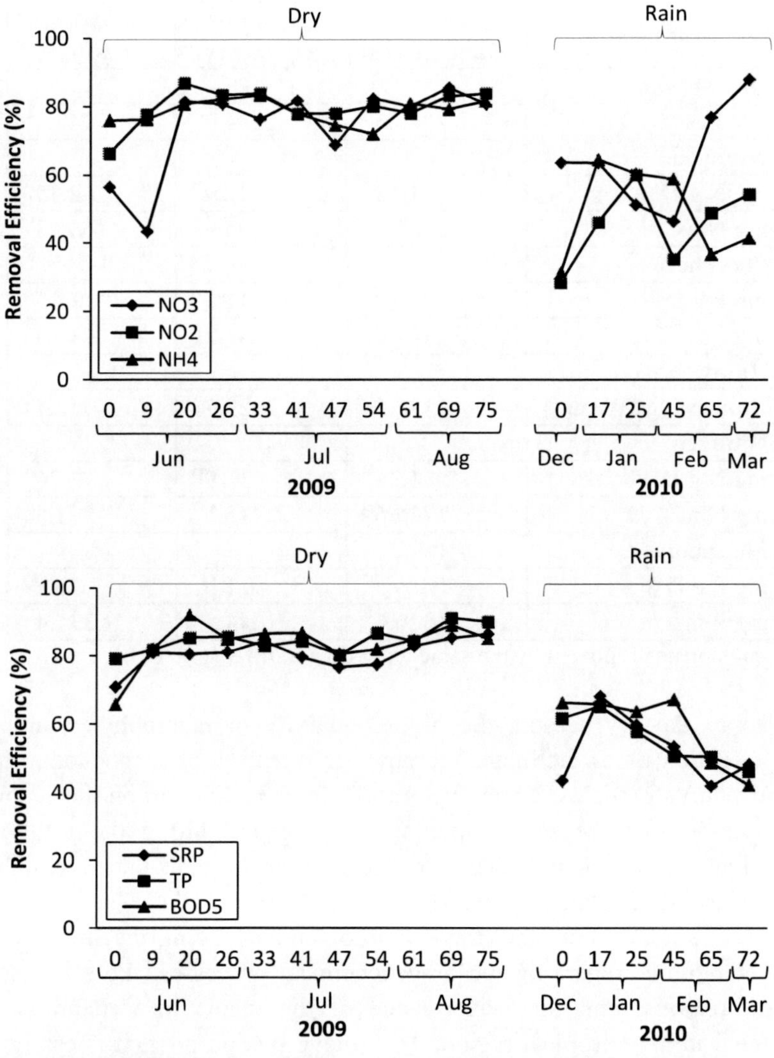

Figure 2. Mass loading removal efficiency of nitrate (NO_3), nitrite (NO_2) and ammonia (NH_4), soluble reactive phosphorus (SRP), total phosphorus (TP) and biochemical oxygen demand (BOD_5) during the dry and rainy seasons.

Table 3. Mean of biomass, nitrogen and phosphorus concentrations in the macrophyte vegetal tissue during the dry and rainy seasons

Biomass	*Typha domingensis*	*Cyperus giganteus*	*Eichhornia crassipes*
Dry Season			
Biomass (g DW m^{-2})	2226.5^{A}	2244.9^{A}	743.4^{B}
Nitrogen			
mg N g DW^{-1}	1.13^{B}	1.29^{B}	2.15^{A}
mg Nm^{-2}	2515.9^{B}	2895.9^{A}	1598.3^{C}
Phosphorus			
mg P g DW^{-1}	0.92^{ns}	0.82^{ns}	0.96^{ns}
mg Pm^{-2}	2044.8^{A}	1848.2^{B}	716.4^{C}
Rainy Season			
Biomass (g DW m^{-2})	1372.5^{B}	2029.7^{A}	352.1^{C}
Nitrogen			
mg N g DW^{-1}	1.25^{B}	1.19^{B}	2.76^{A}
mg N m^{-2}	1715.6^{B}	2415.3^{A}	971.8^{C}
Phosphorus			
mg P g DW^{-1}	0.97^{ns}	0.81^{ns}	0.95^{ns}
mg P m^{-2}	1326.4^{B}	1644.5^{A}	335.4^{C}

In each row, means followed by the same letter do not differ ($p<0.05$).

Also in the dry season, the higher quantity of macrophytes and their longer retention times facilitates the physical retention of suspended nitrogen and phosphorus particles, which were prone to accumulation on the sediment. The observed increase in the macrophyte biomass during the dry season triggered an enhancement in the number of microorganisms that were adhered to the surface of the plants (Table 3).

Current studies show the importance of understanding the limitations in the water regimes regarding the biogeochemical processes, the efficiency of pollutant removal and the performance of the plants in wetland systems constructed in a neotropical region. To achieve this purpose, it is essential to understand how temperature fluctuations between climatic seasons, rainfall, hydraulic loading rate, retention time and growth of plant biomass can help increase the efficiency of the treatments in constructed wetland systems.

The low performance of macrophytes in removing chemical elements during the rainy season may be partially explained by the high hydraulic loading, i.e. the rapid water turnover in the system. Importantly, plants are also influenced by the climate, which makes *C. giganteus, T. domingensis* and *E. crassipes* the most adequate species for the treatment of effluents of constructed wetland systems in southeastern Brazil, as well as in other neotropical regions.

In Brazilian aquaculture farms, the common management practices employed usually build continuous water flow systems in such a manner that water from a given fishpond flows into another fishpond without prior treatment. This cycle can eventually cause a considerable increase of plants and other organic products in the fishpond. In the aquaculture farm of the São Paulo State University, one fishpond (4,268 m^2 in area and 1 - 1.3 m in depth) was totally occupied by macrophytes (Figures 1-D and 3). This fishpond is the fourth of six similarly sized fishponds in a sequential line, and each one of them is directly or indirectly receiving water from the previous one (Figure 1-D).

The initial macrophyte infestation began in 2009 and occupied the entire area of the fishpond. This infestation was carried by the species *Salvinia auriculata* Aublet, which is considered to be an invasive plant. The use of incorrect methods to remove the plants, allied to high temperatures during that period (September 2009) and plenty of nutrients promoted a favorable environment to the proliferation of other invasive species of macrophytes, such as *Panicum repens*, *Typha domingensis*, *Cyperus* sp. and *Polygonum* sp, and the floating species *Salvinia auriculata* Aublet, *Eichhornia crassipes* and *Lemna minor* (Figure 3).

During the rainy season, the lack of macrophytes resulted in a higher presence of autochthonous and allochthonous materials, thus increasing the concentration of organic matter in the fishpond. These factors influenced the decomposition process (BOD_5) and increased the electrical conductivity (Cond) of the water, thus accelerating eutrophication processes (Figure 4). According to Caron & Hutchins (2012), the concentration of organic matter can increase in an exponential manner, thus leading to changes in the zooplankton taxonomic composition and diversity.

The BOD_5 levels and temperature facilitated the growth of Coscinodiscophyceae and inhibited the proliferation of Cladocera. This phenomenon can be partially explained by the high concentrations of suspended solids and organic matter found in the system, which triggered a decrease in the availability of algae that are consumed by Cladocera (Güntzel et al., 2010).

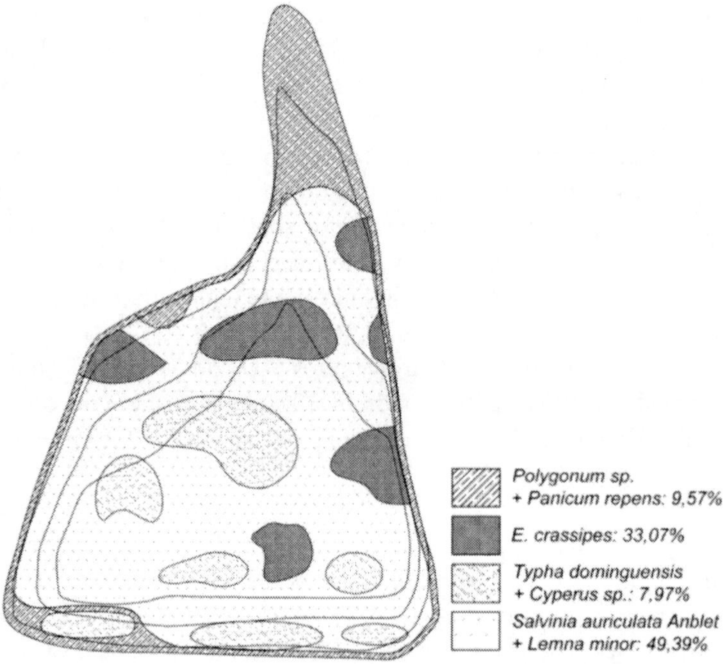

Figure 3. Layout of the total area of the fishpond containing the invasive macrophytes species *Panicum repens*, *Typha domingensis*, *Cyperus* sp., *Polygonum* sp., and floating macrophytes *Salvinia auriculata* Aublet, *Eichhornia crassipes* and *Lemna minor*.

The occurrence of Cyclopoida is commonly associated with eutrophic environments. Eutrophic environments support these filter feeding organisms in capturing Cyanobacteria, which are very common in fishponds (Pinto-Coelho et al., 2005). After the removal of macrophytes, the abundance of Cyclopoida declined, while the abundance of *Aphanocapsa delicatissima* and *Planktolyngbya contorta* increased, indicating a rise of Cyanobacteria in the fishponds (Figure 4).

In the fishponds, higher temperature and increased concentrations of organic material, in conjunction with periods without macrophytes, could affect the abundance of Cyanobacteria species. In freshwater fishponds presenting eutrophic conditions, Cyanobacteria are known to dominate phytoplankton development during the warmest periods of the year (Figure 4). Moreover, the concentration of BOD_5 was lower during the period with macrophytes due to the presence of *E. crassipes* stands. This result is in accordance with Poudel & Simon (2008) statement that mentions the ability of this aquatic plant in reducing the concentration of BOD_5, mostly because its roots are covered with organic material and mucilaginous biofilm. *Lemna minor*, another species present in the studied fishpond, showed excellent potential in reducing the concentration of pollutants such as BOD_5, total suspended solids and thermotolerant coliform (Figure 4).

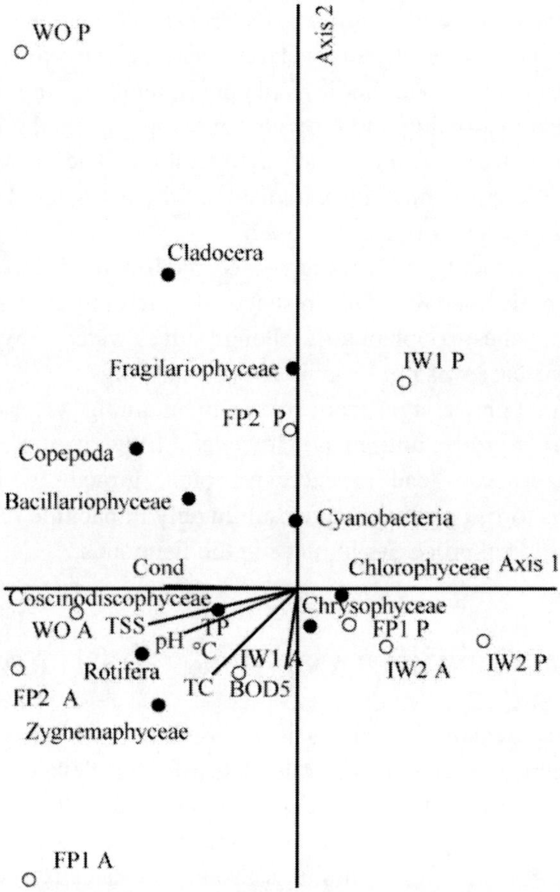

Figure 4. Canonical correspondence analysis (CCA) biplot illustrating the relationships between zooplankton and phytoplankton abundances, and physical and chemical parameters, during two periods, presence (P) and absence of macrophyte (A). Cond = conductivity; °C, temperature; TC, thermotolerant coliforms; TSS = total suspended solids; BOD_5 = 5-day biochemical oxygen demand; WO = water outlet; IW_1 and IW_2 = inlets water; FP_1 and FP_2 = sites inside the fishponds; closed circles = phytoplankton and zooplankton; open circles = sites.

The data also show that the removal of macrophytes and high temperatures caused an increase of the trophic level in the system, which indirectly reflected in the amount of planktonic, bacteria and organic matter present in the fishpond. This eutrophic environment favored particular groups, such as the Rotifera (*Brachionus havanaensis*, *Brachionus calyciflorus*, *Keratella cochlearis* and *Filina longiseta*). Also, an increase of Coscinodiscophyceae and Cyanobacteria was registered, which caused deleterious effects in the water bodies and consequently for the fish.

The presence of floating and emergent macrophytes had a positive effect on the community of zooplankton and phytoplankton. The presence of aquatic plants changed the environmental parameters, which triggered the expansion of major groups of organisms such as Cladocera, Copepoda and Chlorophyceae, providing better sources of natural food for the fish. It is importantly to remember that the presence of macrophytes also favored the development of mesozooplankton, though the water physical-chemical parameters in this case was not suitable for fish farming.

Aquatic plants play a significant role in maintaining water quality, due to their ability to absorb nutrients. However, little control over critical environment issues can lead to excessive plant growth, which can cause serious problems to fish production, and ultimately impact the function and the structure of ecological processes in place in the fishponds.

Macrophytes and Algae Culturing

The growing demand for microalgae, combined with the high costs of medium synthesis has encouraged the research for developing alternative medium. Maximizing production while minimizing costs is the final goal for everyone involved with algae culturing. A traditional approach to identify the factors required for the optimal biomass production usually starts by using a well defined medium, to which the addition of other elements will contribute for the optimization of the nutritive value and consequently, growth of microalgae (Dalay et al., 2007). The need to improve the cultivation of microalgae to be used as feed for farmed fish larvae, pigment in aquaculture and also as supplementary additive in the food industry has triggered many studies in different fields. The ideal alternative medium should increase the production of microalgae while decreasing production costs (Sipaúba-Tavares & Pereira, 2008).

Recent research indicated some alternative solutions such as NPK fertilizer and fertilizers associated with macrophyte and complex B vitamin (Sipaúba-Tavares et al., 2009). NPK fertilizer is a low-cost nitrogen fertilizer that has presented good results in association with microalgae cultivation. The choice of medium used depends thus on several factors, such as the growth requirements of the algae, how the constituents of the medium may affect the final product quality and costs (Dalay et al., 2007). Alternative culture media such as the inorganic fertilizer (NPK, 20-5-20) have been used in our laboratory with great production of *Ankistrodesmus gracilis* (Chlorophyceae) and *Haematococcus pluvialis* (Chlorophyceae) (Table 4).

Another alternative medium used for growing these microalgae consists of a mixture of macrophyte (*Eichhornia crassipes*) extract and NPK fertilizer (20-5-20). The mixture contains approximately 5 kg of *Eichhornia crassipes* that were ground and boiled in distilled water for one hour. The hot extract was filtered and autoclaved at 120°C, during 20 minutes. After autoclaving, a sample (70 mL) was collected and after cooling it was diluted with distilled water, to a total of 1.3 L (Table 4). Additionally, 2.5 ml NPK and 100 ml inoculum of *A. gracilis* and *H. pluvialis* microalgae was added to obtain an approximate density of 3×10^5 cells.ml^{-1} and 0.26×10^5 cells.ml^{-1}, respectively (Sipaúba-Tavares et al., 2009).

The alternative medium NPK (20-5-20) and macrophyte extract was used for growing *A. gracilis* and *H. pluvialis* cultures in a 2-L container submitted to continuous aeration by air and kept at 23 ± 2 °C. The results of this study indicated additional benefits brought by the use of fertilizer and fertilizer with macrophytes as a medium for growing microalgae, since it not only increased the growth rate (Figure 5), but also provided an economic profit. The obtained data demonstrated that the growth rates in both microalgae did not change when different media was used. An exponential growth was observed until the 11th day for the *A. gracilis* and until the 13th day for the *H. pluvialis*. However, the highest cell count (25.2×10^5 cells.mL^{-1}) was obtained with *A. gracilis* in NPK medium, on the 11th day. The density of *A. gracilis* remained above 14×10^5 cells.mL^{-1} in NPK medium on the 11th day, while the *H. pluvialis* density remained below 2×10^5 cells.mL^{-1} (Figure 5). *Haematococcus pluvialis* exhibited smaller growth when compared to other microalgae, due to its complex life cycle (motile and non-motile cells and resting cell or cyst). The alternative media used were satisfactory for maintenance of cultures containing high-density of *H. pluvialis* (Figure 5).

Table 4. Nutrient composition of two alternative media (NPK and Macrophyte + NPK) used in cultivation of microalgae

Constituents	NPK ($g.L^{-1}$)	Macrophyte + NPK ($g.L^{-1}$)
P_2O_5	2.5	-
K_2O	10	-
N	10	0.095
P	-	0.091
C	-	0.199
Mn	-	0.06
Mg	-	0.003
Fe	-	2.13
K	-	0.056
Ca	-	0.001
Cu	-	0.05 mg
Thiamine (B_1)	0.007	0.007
Vit B_2	0.007	0.007
Vit B_6	0.005	0.005
Vit B_{12}	33 µg	33 µg
Vit H	0.01 mg	0.01 mg

* The Composition of macrophyte extract (*E. crassipes*) plus NPK fertilizer was based according Sipáuba-Tavares et al. (2009).

Using an inorganic fertilizer based medium (NPK, 20-5-20), in association with macrophyte (*E. crassipes*), we obtained a new medium formulation that proved to be efficient, at least in batch culture mode, promoting high-density microalgae cultures. Thus, NPK (20-5-20) may be a good tool for maintaining *H. pluvialis* in the fishpond.

Microalgae culture is currently a viable commercial source of nutrients, which yield a high biomass and nutritional value; however the success of production is still not a consensus. Further studies will be necessary in order to develop an ideal alternative medium to maintenance of high-density cultures of different Chlorophyceae species. Importantly, the technology using macrophyte and inorganic fertilizer can be used as a good alternative of microalgae culture due to its low cost and for showing good results in the growth of *A. gracilis* and *H. pluvialis*.

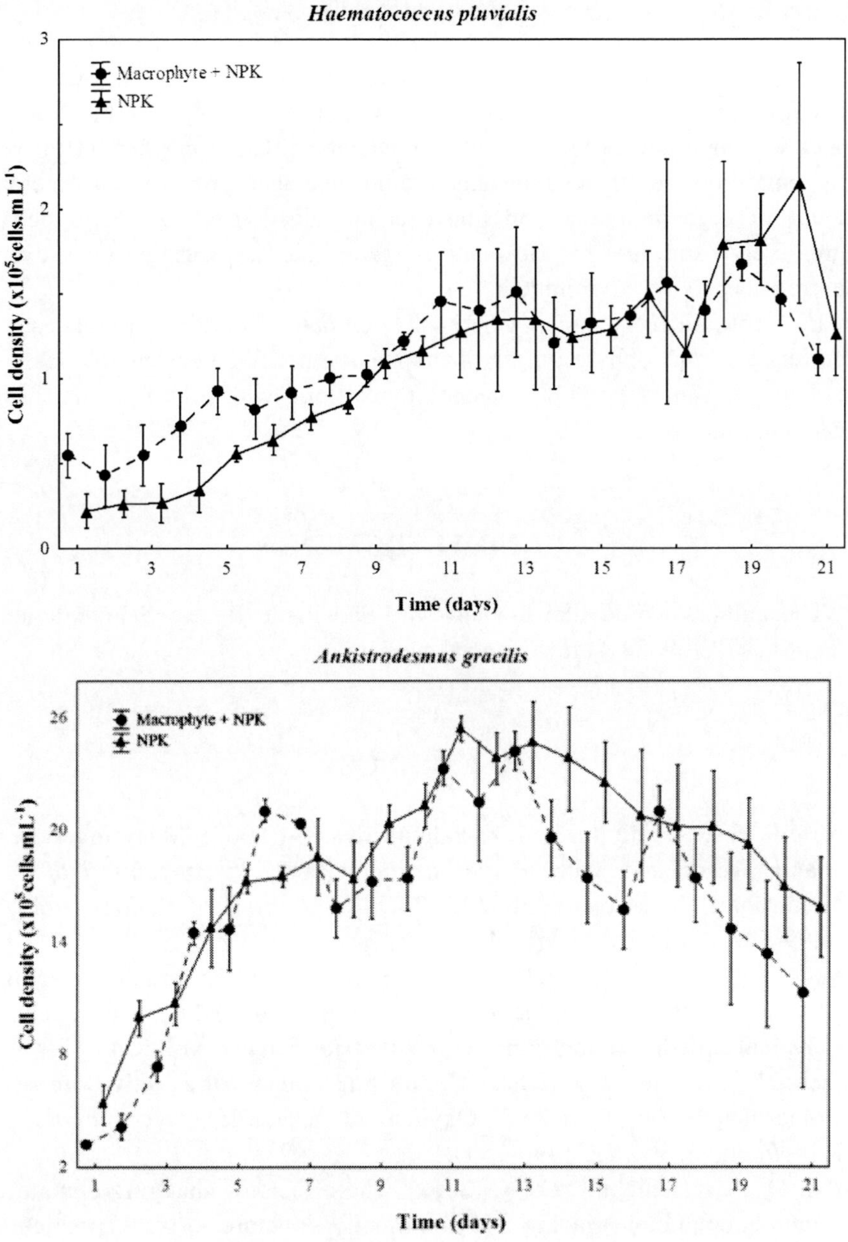

Figure 5. Growth of *Haematococcus pluvialis* and *Ankistrodesmus gracilis* in NPK (20-5-20) and NPK + macrophyte media, in the laboratory.

Conclusion

Macrophytes can be well managed to act as tools of high success in aquaculture, in the treatment of effluents (wetland), water supply (bio-filter) or as a culture medium for growth and maintenance of Chlorophyceae. However, if the conditions in artificial systems are not managed properly, an excessive amount of organic matter and nutrients may lead these plants to behave aggressively and cover the entire water surface, triggering adverse consequences to the environment.

In neotropical regions, where the climate naturally stimulates the occurrence of macrophytes, these plants may be used in a sustainable manner, thus becoming an important biotechnological tool, with high potential in the aquaculture industry.

Acknowledgments

The authors would like to thank the São Paulo Research Foundation - FAPESP (2012/09884-4) for financial support.

References

Albertoni, E. F., Prellvitz, L. J. & Palma-Silva, C. (2007). Macroinvertebrate fauna associated with *Pistia stratiotes* and *Nymphoides indica* in subtropical lakes (south Brazil). *Brazilian Journal of Biology*, vol. *67*, 499-507.

Boyd, C. E. & Queiroz, J. (2001). Feasibility of retention structures, settling basins, and best management practices in effluent regulation for Alabama channel catfish farming. *Reviews in Fisheries Science,* vol. *9*, 43-67.

Caraco, N. F. & Cole, J. J. (2002). Contrasting Impacts of a Native and Alien Macrophyte on Dissolved Oxygen in a Large River. *Ecological Applications*, vol. *12*, 1496-1509.

Caron, D. A. & Hutchins, D. A. (2012). The effects of changing climate on microzooplankton grazing and community structure: drivers, predictions and knowledge gaps. *Journal of Plankton Research*, vol. *1*, 1–18.

Chen, H. J., Zamorano, M. F. & Ivanoff, D. (2010). Effect of flooding depth on growth, biomass, photosynthesis, and chlorophyll fluorescence of *Typha domingensis*. *Wetlands*, vol. *30*, 957-965.

Christie, H., Norderhaug, K. M. & Fredriksen, S. (2009). Macrophytes as habitat for fauna. *Marine Ecology Progress Series*, vol. *369*, 221-233.

Dahms, H. U., Hagiwara, A. & Lee, J. S. (2011). Ecotoxiology, ecophysiology, and mechanistic studied with rotifers. *Aquatic Toxicology*, vol.*101*, 1-12.

Dalay, M. C., Imamoglu, E. & Demirel, Z. (2007). Agricultural fertilizers as economical alternative for cultivation of *Haematococcus pluvialis*. *Journal of Microbiology and Biotechnology*, vol. *17*, 393–397.

Dias, S. G. & Sipaúba-Tavares, L. H. (2012). Physical, chemical and microbiological aspects during the day and rainy seasons in a pond covered by macrophytes used in aquaculture water supply. *Acta Limnologica Brasiliensia*, vol. *24*, 276-284.

Eler, M. N., Espíndola, E. G., Espíndola, E. A., Brigante, J., Nogueira, M. M., Nogueira, A. M. & Millani, T. J. 2006. Fee-fishing farm water quality and sediment assessment: Physical, chemical and biological analyses and toxicity bioassays. In: Eler, M.N., Espíndola, E.G. (ed.) Fee-fishing farm impacts: Analyzed activity in hydrographic basin of Mogi-Guaçuriver. Rima, São Carlos-SP, Brazil., 101-144. (Portuguese)

Ferdoushi, Z., Haque, F., Khan, S. & Haque, M. (2008). The effects of two aquatic floating macrophytes (*Lemna* and *Azolla*) as biofilters of nitrogen and phosphate in fish ponds. *Turkish Journal of Fisheries and Aquatic Sciences*, vol. *8*, 253-258.

Güntzel, A. M., Panarelli, E. A., Silva, W. M. & Roche, K. F. (2010). Influence of connectivity on Cladocera diversity in oxbow lakes in the Taquari River floodplain (MS, Brazil). *Acta Limnologica Brasiliensia*, vol. *22*, 93-101.

Henry-Silva, G. G. & Camargo, A. F. M. (2006). Efficiency of aquatic macrophytes to treat Nile tilapia pond effluents. *Scientia Agricola*, vol. *63*, 433-438.

Henry-Silva, G. G., Camargo, A. F. M. & Pezzato, M. M. (2008). Growth of free-floating aquatic macrophytes in different concentrations of nutrients. *Hydrobiology*, vol. *610*, 153-160.

Kadlec, R. H. & Wallace, S. D. (2009). Treatment wetlands. Boca Raton: Taylor and Francis Group. 1020p.

Kaplan, D., Bachelin, M., Muñoz-Carpena, R. & Chacón, W. R. (2011). Hydrological importance and water quality treatment potential of a small freshwater wetland in the humid tropics of Costa Rica. *Wetlands*, vol.*31*, 1117–1130.

Macedo, C. F. & Sipaúba-Tavares, L. H. (2005). Variações de nutrientes e estado trófico em viveiros sequenciais de criação de peixes. *Acta Scientiarum Animal Science*, vol. *27*, 405-411.

Maltais-Landry, G., Maranger, R., Brisson, J. & Chazarenc, F. (2009). Nitrogen transformations and retention in planted and artificially aerated constructed wetlands. *Water Research*, vol. *43*, 535-545.

Millan, R. N., Sipaúba-Tavares, L. H. & Travaini-Lima, F. (2014). Influence of constructed wetland and soil filter systems in the dynamics of phytoplankton functional groups of two subtropical fish farm wastewaters. *Journal of Water Research and Protection*, vol. *6*, 8-15.

Onsem, S. V., Backer, S. & Triest, L. (2010). Microhabitat-zooplankton relationship in extensive macrophyte vegetations of eutrophic clear-water ponds. *Hydrobiologia*, vol. *656*, 67-81.

Perbiche-Neves, G., Ferrareze, M. F., Serafim-Jr, M., Shirata, M. T. & Lagos, P. E. D. (2011). Influence of atypical pluviosity on phytoplankton assemblages in a stretch of a large sub-tropical river (Brazil). *Biologia*, vol. *66*, 22-41.

Pinto-Coelho, R. M., Bezerra-Neto, J. R. & Morais-Junior, C. A. (2005). Effects of eutrophication on size and biomass of crustacean zooplankton in a tropical reservoir. *Brazilian Journal of Biology*, vol. *65*, 325-338.

Poudel, D. D. & Simon, M. J. (2008). Pasture-based dairy and water hyacinth (*Eichhornia crassipes*) for reduced biological oxygen demand in dairy discharge water. *Outlook on Agriculture*, vol. *37*, 135-142.

Sandilands, K. A. & Hann, B. J. (1997). Is phytophilous zooplankton community structure affected by nutrients and fathead minnows? *UFS (Delta Marsh) Annual Report*, vol. *32*, 47-54.

Santiago, A. F., Calijuri, M. L. & Luis, P. G. (2005). Potential for the use of wetlands systems in wastewater treatment: A contribution to sustainability of water resources in Brazil. *Natureza & Desenvolvimento*, vol. *1*, 29-39. (Portuguese)

Sipaúba-Tavares, L. H., Favero, E. G. P. & Braga, F. M. S., (2002). Utilization of macrophyte biofilter in effluent from aquaculture: I. Floating plant. *Brazilian Journal of Biology*, vol. *62*, 713-723.

Sipaúba-Tavares, L. H., Barros, A. F. & Braga, F. M. S. (2003). Effect of floating macrophyte cover on the water quality in fishpond. Acta Scientiarum: *Biological Sciences*, vol. *25*, 101-106.

Sipaúba-Tavares, L. H. & Pereira, A. M. L. (2008). Large scale laboratory cultures of *Ankistrodesmus gracilis* (Reisch) Korsikov (Chlorophyta) and *Diaphanosoma birgei* Korinek, 1981 (Cladocera). *Brazilian Journal of Biology*, vol. *68*, 875-883.

Sipaúba-Tavares, L. H., Ibarra, L. C. & Fioresi, T. B. (2009). *Ankistrodesmus gracilis* (Reisch) Korsikov Chlorophyta) laboratory cultured in CHU_{12} and macrophyte whith NPK media. *Boletim do Instituto de Pesca*, vol. *35*, 111-118. (Portuguese)

Sipaúba-Tavares, L. H., Lourenço, E. M. & Braga, F. M. S. (2010). Water quality in six sequentially disposed fish ponds with continuous water flow. *Acta Scientiarum: Biological Sciences*, vol. *32*, 9-15.

Souza, J. T., Van Haandel, A., Lima, E. P. C. & Henrique, I. N. (2004). Use of constructed wetland for the post-treatment of domestic sewage anaerobic effluent from UASB reactor. *Engenharia Sanitária Ambiental*, vol. *9*, 285-290.

Travaini-Lima, F. & Sipaúba-Tavares, L. H. (2012). Efficiency of a constructed wetland for wastewaters treatment. *Acta Limnologia Brasiliensia*, vol. *24*, 255-265.

Ulrich, H., Klaus, D., Irmgard, F., Annette, H., Juan, L. P. & Regine, S. (2005). Microbiological investigations for sanitary assessment of wastewater treated in constructed wetlands. *Water Research*, vol. *39*, 4849–4858.

Van Dam, A., Beveridge, M. C. M., Azim, M. E. & Verdegem, M. C. J. (2002). The potential of fish production based on periphyton. Reviews in *Fish Biology and Fisheries*, vol. *2*, 1-31.

Zhang, Z., Rengel, Z., Liaghati, T., Antoniette, T. & Meney, K. (2011). Influence of plant species and submerged zone with carbon addition on nutrient removal in storm water bio-filter. *Ecological Engineering*, vol. *37*, 1883-1941.

INDEX

C

D

E

T

U

V

W

Y

Z